TRAMWAY TITAN

BYRON RIBLET,
WIRE ROPE AND WESTERN
RESOURCE TOWNS

MARTIN J. WELLS

Order this book online at www.trafford.com
or email orders@trafford.com

Most Trafford titles are also available at major online book retailers.

© Copyright 2011 Martin J. Wells.
All rights reserved. No part of this publication may be reproduced, stored in a retrieval system, or transmitted, in any form or by any means, electronic, mechanical, photocopying, recording, or otherwise, without the written prior permission of the author.

Printed in the United States of America.

ISBN: 978-1-4120-5093-7 (sc)

Trafford rev. 08/02/2011

 www.trafford.com

North America & International
toll-free: 1 888 232 4444 (USA & Canada)
phone: 250 383 6864 ♦ fax: 812 355 4082

Dedication

To my parents, Edward and June, who showed me the value of books and who first took me to the Lardeau Region.

Acknowledgments

Appreciation needs to be expressed to many people-to both the Nelson and Salmo Archives a thank you is in order for their warm and kind support; to the Vancouver Public Library for their invaluable and unexcelled help and assistance navigating their wonderful collection; huge accolades are forwarded to Robert McLellan for his patience, knowledge and sharing of his authoritative collection; to Bill Laux-compiler, compatriot and co-pilot a nod is needed; also, Jennifer Sotelo of Colorado School of Mines was helpful in locating materials; Dr. Jeremy Mouat, History Professor at Athabaska University for his insights; Jim Ellis of Aerial Engineering for his work experience and contacts at Riblet Tramway; John Fahey for his prior research; Milford Homer of the New Mexico Department of Mineral Resources; Logan Hovis and the US Parks Service, both in Seattle and Kennecott Alaska; Joe Kurtak of the Department of the Interior; Bob Tapp of the Hollyburn Historical Association; Rayette Wilder of the Eastern Washington Historical Society Archives was very efficient; Lora Feucht of the Lewiston Museum and the Land of the Yankee Fork Visitor Centre in Challis, Idaho were most welcome and supportive. Charles Riblet should also be thanked for his sharing of information. While Murray Lundberg, Candy Moulton, Milton Parent, Don Blake, D. Wilkie and R. Turner, Garnet Basque, Robert Trennert, John Davis and D.M. Wilson should also be thanked for their printed works. Fulcrum Publishing kindly granted permission for use of the excerpt from Teri Hein's *Atomic Farmgirl* on page 32. Eileen Mak and Meg Stanley were kind in their sharing of historical methods and library card. Michael Carter of the BC Archives and Records Service lent his time. Appreciation is also due to ZESS, and CCS, Doug Lemmon, and Bill Dietrich, the Lovestrom Family, the Shanks Family, and Moore Family, the Southern Family; the Hoffman family; the Froslevs of Brackendale, the Summerfield family; S. Laurence and B. Porter for systems support. Also to be included are N. Brady-Browne, R. Roberts, the CSPIAA, BHS and the ATA for engineering explanations and solutions; and finally to Chuck Dwyer—late of the US Forest Service—for pointing me in the right direction. The editing was kindly performed by Shannon Wilson, Patricia Wilson and Toni Thompson; design and typesetting by Larry M. Farnand. This work would not have been able without the patience of my family.

Contents

Introduction	**1**
1. Backgroung History, Andrew Hallidie and the Hall Mine	**13**
1. Background History	15
2. Andrew Hallidie	24
3. Silver King	28
4. Pullman and Tekoa, Washington	32
5. Streetcars in Spokane	35
6. Washington Water andPower Company, Spokane	35
7. The Riblet Family	37
Notes—1	40
2. Cableways in the Kootenays	**47**
8. Sandon, BC	49
9. Silverton, BC	64
10. Moyie, BC	68
11. Ymir, BC	71
12. Porto Rico, BC	74
Notes—2	75
3. Technical Details and the Trenton Iron Works	**79**
13. Techinical Details of Mining Aerial Trams	81
14. Competition	89
15. A. Leschen and Sons Aerial Tramway Company	91
Notes—3	93
4. The Lardeau and the Kootenay Lake Mining Districts	**95**
16. The Lardeau District	97
17. Southern Kootenay Lake Mines	115
Notes—4	119
5. South West Coast	**123**
18. Mount Sicker and the Tyee Mine	125
19. Britannia Beach BC	129
Notes—5	134
6. The Boundary Region	**135**
20. Grand Forks, BC	137
21. Phoenix, BC	137
22. Salmo, BC	140
23. Blakeburn, BC	142
Notes—6	149
7. People and Farther Places	**153**
24. Royal Newton Riblet	155
25. Encampment, Wyoming	158
Notes—7	163

8. The North: Alaska and the Yukon in the Edwardian Era 165
    26. Conrad City, Yukon 167
    27. Akun Island, Alaska.................................. 172
    28. Ketchikan District 173
    29. Prince William Sound, Alaska 174
    30. Dolly Varden 175
        Notes—8 177
9. Enter the Guggenheims 179
    31. The Guggenheims 181
    32. Braden Copper Company, Chile 187
    33. Stewart, BC 188
        Notes—9 201
10. Washington State 203
    34. Chewelah, Washington 205
    35. Metaline Falls, Washington 207
    36. Ione, Washington 207
    37. Sumas and Concrete, Washington.................. 209
    38. The Snake River Trams 211
        Notes—10 213
11. The U.S. West 215
    39. Oatman, Arizona 217
    40. Yosemite, California............................... 218
    41. Jarbridge, Nevada 218
    42. Terarro, New Mexico 218
    43. American Fork, Utah 223
        Notes—11 224
12. South America 227
    44. Bolivia 229
    45. Peru 231
        Notes—12 233
13. The Depression 235
    46. The Depression 237
    47. Windpass Mining, BC 239
    48. Hedley, BC 239
    49. Mt. Hood, Oregon 241
        Notes—13 246
14. War and its Aftermath 249
    50. Bishop, California 251
    51.-Mouat, Montana 253
    52. Yugoslavia 254
    53. Albania 258
    54. Nalaahu, Hawaii 261
        Notes—14 262

15. Natural Resources and New Recreations 265
    55. Hollyburn Ridge, Vancouver, BC 267
    56. Cassiar Asbestos and the Clinton Creek Mine 268
    57. Nepal 269
        Notes—15 272

Conclusion 273

Afterword 275

APPENDICES 279
    APPENDIX I-Coastal BC Tramways 280
    APPENDIX II-Patents of Byron Riblet 282
    APPENDIX III-Other Cable Systems 283
    APPENDIX IV-Rexspar Uranium Mines 290
    APPENDIX V-Riblet Ski Chairlifts (1953-1969) 298
    APPENDIX VI-Aerial Tramways on the Chilkoot Pass 303
    APPENDIX VII-Labour Leaders and Mining 306
    APPENDIX VIII-Colombia 309
    APPENDIX IX-Asia 310
    APPENDIX X-Cableways 312

Glossary 314

Maps
    British Columbia VIII
    Pacific Northwest IX
    Alaska X
    Sandon 52
    Silverton 65
    Lardeau 98
    Conrad 169
    Stewart 190

Bibliography 316

Index 331

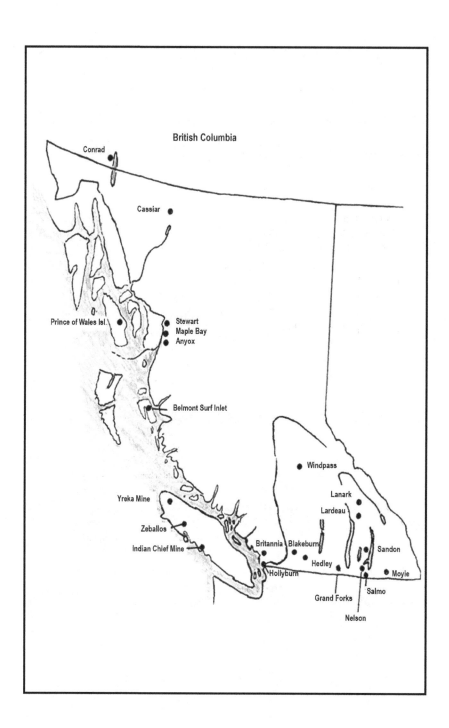

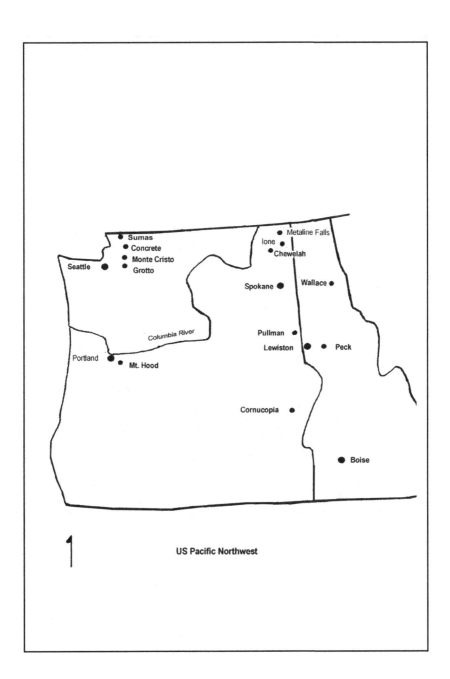

US Pacific Northwest

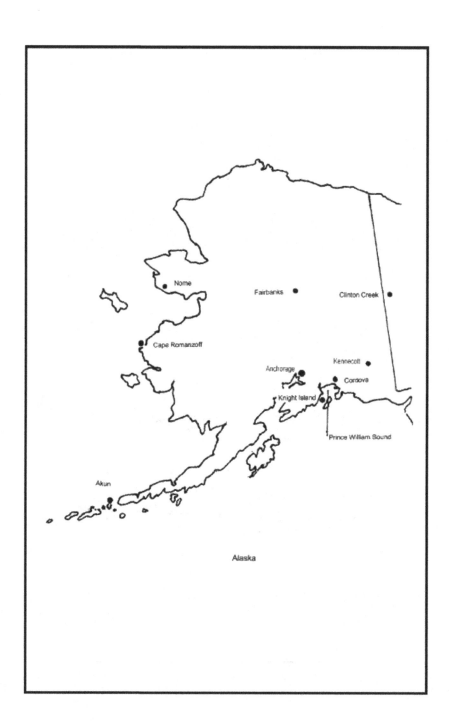

Introduction

Byron Riblet's Contributions to the Mining Industry

British Columbia's stunningly beautiful mountains have inspired some similarly impressive feats of engineering by mankind. It was the geology of those mountains, in the form of mineral wealth, which created a need for structures which would allow access to the peaks. As a result, British Columbia became a cradle of creation for a host of large mining cable systems. That province saw the development of a highly successful line of mining tramways, tramways which would gain greater fame through use in mountain regions around the world.

Byron Riblet (1865-1952) was a civil engineer who came west from Iowa with the Northern Pacific Railroad and made a name for himself in mine haulage throughout northwestern North America. Riblet was responsible for most of the mining cable systems in and around British Columbia. His firm, the Riblet Tramway Company, erected systems in the Kootenays, the Boundary region, and on the Coast, as well as in the Klondike and the Rocky Mountains. If there was a tramway to be built in the Canadian West or beyond—indeed, as far away as the Andes mountain range—Riblet stood ready to build it.

Some Riblet cableways were ten, twenty or even thirty miles long. The tramways were large, complex structures which pushed the limits of the engineering technology of the day. Yet very little is written about these systems and the remarkable man behind their development.

Initial attempts to examine the history behind, or the details regarding, Riblet's many innovations led nowhere. This discovery prompted a long search by the author, one which took him through the stacks of several public, community college, and university libraries. Mining books, regional histories, and general-interest "Canadiana" works were consulted as well. But the results of this research only served to reveal a large gap in the scholarship that has been brought to bear upon this subject to date.

It is clearly time for a full, definitive history of Byron Riblet and his activities. A chronological work examining his life and successive cable system achievements might have sufficed—a Riblet tramway travel guide, if you will. Indeed, this book does offer a detailed look at Riblet the innovator and his accomplishments, and equally importantly, it attempts to present both within the context of their time, a time so very different from our own. To that end the book discusses at times the prevailing socio-political, economic, and cultural climate of the period in which Riblet blazed his singular trail. As it happens, "the Riblet connection" itself provides a unique entry into, and link among, the many distinct mining regions in which Riblet's tramway systems were employed. Yet, it may correctly be said that the purpose of this work is, in fact, two-fold: in addition to shedding light on one of the most influential and least known, inventors in North American history, this volume addresses the critical importance of mining as science, business, livelihood, and lifeline. Indeed, mining, strictly as a human activity in itself, shares with Riblet the peculiar distinction of having improved the fortunes of humankind immeasurably—while remaining largely forgotten in the present day.

As it was with the wealth of resources once hidden within British Columbian rock and soil, citizens of that province now stand upon a very interesting history. One has only to scratch the surface for that multi-faceted treasure to be revealed.

One first surprising discovery might be that Byron Riblet was not only a pioneer in the mining field, but that he inadvertently came to play midwife to the burgeoning skiing industry. As a sideline to his mining business, Riblet erected some of the very first chairlifts in North America. The first all-steel chair was built by Riblet at Mount Hood, Oregon, in 1938. His company later built the first steel ski chairlift in British Columbia at Hollyburn Ridge in 1952. Eventually this mining engineer's firm went on to become the largest ski chairlift builder in the world by the 1970's.

Before focusing attention on Riblet, however, we must explore mining itself, and its effects on history and the historical process, for metals and mining are an integral part of human history.

The Human Need and Desire for Metals: both Precious and Practical

Entire historic periods have carried the names of metals, such as "The Bronze Age" and "The Iron Age." Without metals our society would not have developed to the degree, or at the rate, it did in the past, nor would it have continued to flourish through the Space Age or the Computer Age. In human society metals carry value beyond the physical; values which spills over into the realm of thought and emotion. Consider the use of gold, silver, and bronze for Olympic medals, or expressions such as "a golden opportunity," which use that element as a positive symbol.

Too, metal-related concepts and expressions sometimes come to us steeped in the symbolism of war: "iron-clad," "armour-plated," and "cold steel" are but a few such terms. Indeed, war itself is usually fought over geography or economics—gold mines, and the greater ability to convert materials into gold, are often the cause for war. War costs money, and the ability to wage it is secured by a nation's holdings of gold and silver. Yet, too, iron is needed for cannon and shot, lead for bullets, and copper for buttons and ships' bottoms.

Strangely, if soldiers are skillful enough with the lethal delivery one nation's metals to the other side, while avoiding flying metal from all sides, then that nation will ultimately award small pieces of brass or silver to certain of its troops in the form of medals. What a peculiar bargain it is—and in the process, the victorious nation usually accumulates more gold and silver still, in the form of tribute.

Such is the value of metals, whether it be gold, grapeshot, or the gallium in our computers. Yet the efforts and advances of the people who deliver such elements up for use are rarely discussed.

Mining at the Heart of Many Disciplines and Developments

Mining is one of those applied sciences which is heavily cross-fertilized. In a way, the ideas involved are like ore bodies, in that while these ideas are connected to a given field of study, they bear shoots that wind in all directions. Some disappear back into the host rock, while others connect to new fields. For example, instead of treating Byron Riblet and his aerial trams, this very work could easily have investigated herein the aggregate decrease in world ore content over the past century, or the demise of the Canadian mining industry as a whole. Or we could have documented the rise of militant labour relations such as those involving the Western Federation of Miners and their vicious battles in Idaho and Colorado.

We might even have treated mining from a more strongly political and economic angle, as a subset of national and international interests. For instance, we may have examined how Byron Riblet and chief competitor Andrew Hallidie's companies grew in the American Empire and its vassal states of Canada, and South America; how Adolf Bleichert and others in the field grew in Mitteleuropa and Russia; and how competitors grew in the British Empire, Africa, South America, and Australasia. Indeed, we might have studied how both Japanese and Soviet tramway designers came eventually to flourish in their own protected regions.

Finally, these pages might have posed the more philosophical question of whether mining technology ultimately benefited its pioneer communities, or whether aerial tramways became in fact the steel tentacles of the Morgan-Guggenheim financial octopus, with its mines, railways and smelters just another element of these men's smothering monopolies.

All of these ideas have mining as their jumping-off point. But again, our own main area of focus will involve one man and his enduring impact on an industry.

Mining in the Americas-A Compact History

Seeking the fastest ocean route to Asia, European sea traders discovered the continents of North and South America. There they also discovered rich mineral deposits. The subsequent Spanish domination of Peru and Mexico yielded huge quantities of silver and gold. An "El Dorado" had been found. A single mine in Bolivia yielded over half the world's silver supply for the entire century following the Spanish conquest of these areas.

But age-old methods of hand labour continued to be used in the mines. Arduous techniques, relatively unchanged since Roman times—chiefly the hand-breaking of rocks, with the application of fire and water where needed—were steadily employed to extract precious metals from the earth. The reduction of ore was still carried out using mule-powered whims, or capstans, or a similarly rudimentary apparatus called the arrastra, which operated on the same principle of slow, grinding, body-driven rotation. Finally, those great volumes of ore collected were sacked in leather pouches for transport on mules or horses.

Such mining continued from the sixteenth century onward. Predictably, problems arose with deeper mines, reduced ore content in given mine, or played-out seams in such a mine. Silver production dropped exponentially with the decades until the latter half of the nineteenth century, when better technology prevailed.

With such labour-intensive processes being used before the improvement of relevant technology, only the most accessible and abundant ore bodies were developed. The identification, development, and production of remote or more mountainous deposits would have to wait for later centuries and the age of machines.

In North America, mining exploded among the placer fields at Sutter's Mill, California in 1848. The ensuing placer finds in Oregon, Idaho, British Columbia, and finally the Klondike, sent miners fanning out throughout the West. Afterward came the silver boom and the establishment of underground mining in Nevada, Montana, and Idaho. Following that came the development of the base metals industries, with copper and lead yielding huge incomes for its operators. In addition, the passing of the Homestead Act (1862) and the

Mining Act (1872) gave free title of land held by the US government, and ownership of western lands was thus transferred to settlers and miners. This European westward expansion was furthered by technology.

In the late nineteenth century, the rise of technology such as the telephone, the telegraph, railways, electrical appliances, and indoor plumbing created markets for base metals. So did the rise of the steam-powered steel ships. The amount of metal that went into the building of the *RMS Titanic*, for example, was in the thousands of tons; metal was needed for the ship's plates, pipes, boilers, engines, bearings, fittings, table settings, and watertight doors. And the Titanic was just one ship among hundreds built during the period; consider the coasters, tramp steamers, and warships all constructed around the turn of the century.

The nascent automobile industry, too, spurred metal production. Tinplate was needed for auto bodies, brass for radiators, and nickel and chrome for plating. And in another developing industry, George Eastman's Kodak camera established one more market for silver.

The base metal boom was curtailed in 1919 with the end of the First World War. The war first created a huge demand for brass shell casings (hence the need for zinc, copper, tin, and lead) but with victory the metals were no longer needed in volume. Mines closed as a result. Old mines, too, closed as their ore bodies played out. But other mines would eventually come online, as better haulage and extraction technology at last allowed for the exploitation of lower-ore-content deposits such as those in the mines at Tererro, New Mexico, and Pleasant Grove, Utah.

The Great Depression returned the lustre to gold and gold mining, as the market became price-stable. Yet this expansion was cut short by the closing of the gold mines during the Second World War. That war had indeed spurred mining of a number of strategic war materials, such as nickel, mercury, magnesium, and vanadium. And in time, the Cold War would create a boom in uranium mining. Clearly, metal markets ride the waves of both technology and world events.

It is the lure of easy profits that draws people to mining. Indeed many of the "penny stocks" listed on the Vancouver Stock Exchange or the Toronto Stock Exchange are issued purely for speculation and not production. No player is completely immune from the "irrational exuberance" which markets promote. The pages of history are littered with the schemes of stock promoters. The most notorious and remembered was the recent Bre-X swindle. It was a Calgary 'penny stock' gold company whose stock value went out of proportion to common sense and its ensuing crash created effects which could be felt throughout North America.

Thus do today's con-men posing as businessmen, in the manner of frontier Colorado's Jefferson "Soapy" Smith, continue to ply their trade—only now their actual shell games, as one may see, are more likely to be run out of Timbuktu or Borneo, not a gangster hideout called Skagway.

Yet with a crashing market often comes the "Next Big Thing." In our day, plastics, electronics, biotech and other technology-stocks all paint pictures of a trouble-free world. So, too, the thinking went in the nineteenth century, only then, the vehicle of human deliverance was to be the railroad. Most people of the time were unaware of the fact that along with the development of railroads, came some diabolical double-entry bookkeeping. To build the railroads required vast quantities of capital, and to get that capital, they had to secure loans, land, or both from the government. Railroads were often not profit-making ventures. In order to drum up revenue to operate, and to repay the loans, one needed the support of settlers, as well as of agricultural, logging, and mining interests. It was crucial that these interests could be counted on to ship out their raw materials and import their manufactured goods from the industrial Great Lakes region via the rails.

With every new railroad winding into mountainous regions came prospectors. Railroad surveyors would often scout for deposits of minerals as they worked to determine routes and possible freight traffic. As a result, the Union Pacific Railroad benefited from locating copper and coal in Utah; the Northern Pacific, silver and coal in Idaho and Washington; and the Canadian Pacific, silver and coal in the Rockies.

The Uses of Aerial Tramways Around the World

The story of aerial tramways in North America in the nineteenth and twentieth centuries mirrors the pattern of European expansion westward. Old ideas were transplanted to a new region, and were adapted in relation to local conditions. Those local conditions fostered further technical development which, in turn, promoted whole new industries. The story of aerial tramways is an integral part of the expansion and settling of the West. For this reason, mining history itself is connected to a wide range of other areas.

Aerial tramways turn up in a variety of places, some quite surprising. The utility of the design and its transferability is truly amazing. For instance, banana bunches to this day are moved by small tramways on the plantations of Costa Rica. Tea in India's Assam state continues to be moved from hillsides to transfer points by aerial tramway.

Just about everything seemed, at one time, to have been transported by tramway: coal in Northumbria and Pennsylvania; pulpwood in Quebec; Air Force personnel to a bleak Distant Early Warning site on the Bering Strait; freight through the Chilkoot Pass on the British Columbia-Alaska border; cement in Washington state; wheat in Idaho; lumber in Northern California; borax in Death Valley; sugar in Hawaii; coffee in Colombia; mahogany in Haiti; trucks on the Tigris River; and Italian munitions at Isonzo and Ethiopia.

Yet the bulk of the work of aerial trams remained with metal mines. The ground covered by mining tramways was and is staggering. Almost every country had operations, from Norway to South Africa, from the Soviet Union to Australia; from Canada to Argentina; and, without exaggeration, virtually all points in between, spanning the continents.

Prior to the Great Depression, the number of tramway manufacturing companies was in the double-digits. These companies produced hundreds of trams. While the Depression certainly contributed to the decline of cable tramway building, there was, in addition, the corresponding emergence of more efficient technologies such as trucking. The mining industry soon graduated to large open-cast operations, essentially leaving underground excavation behind.

Modern Advances Originating From Our Mining Past

The use of aerial trams in mining may be considered symbolic of the subtle, ongoing changes in that industry: from hand labour to high-tech computers; from pack-animals to subsea exploration; from surface deposits to satellite mapping, automation, and underground cell phones. Today, increased mining production corresponds with a twenty-year pattern of falling metal prices, coupled with the technological obsolescence of copper in the telephone industries due to the advent of glass fibre-optic methods.

The days of transporting extracted ore and equipment by mule are long gone. Materials are now shifted with huge, 200-ton heavy-haul trucks. In addition, miners may now slurry ore in rubber pipes, or simply refine it on-site through heap leaching.

In each of the market booms, miners never feared trying new technology and allowing inventors the room to experiment and perfect. Aerial tramways are only one perfection; others include helicopters (the ultimate aerial tram), shot-creting tunnels, rock drills, conveyor belts, magnetometers, computer modeling, remotely-operated vehicles, and satellite phones.

A development stemming directly from our mining past is the increase in the general use of steel wire in modern society. Cable tramways are but one indicator of the dominance of steel cable use in our daily lives. We drive over bridges suspended by cables, the elevators that lead to our offices are whisked into the air by cables, our power lines are held aloft with cables and cement bridges and buildings are pre-stressed with cable; our electric dams, gravel pits, and logging operations all depend on cable systems. Subsea telephone lines are shielded with steel cable. Cables are the ligaments that join and move the segments of our society. Indeed, without cables access to the more remote regions would be impossible. One of the more out of the way places is Bolivia, and it has had a curious history.

Bolivia—The Rise and Fall of a Mining Nation

To understand the depth to which a region's fate is bound to its mining heritage, one need only consider the peculiar fate of Bolivia. It once was a very rich state. Bolivia boasted fabulous reserves of silver ore from the time of the Spanish conquest onward; some of this very silver was minted into the Spanish Main's "pieces of eight" that were traded in China, by way of Mexico and the Philippines, for porcelain, silk and spice. This silver also financed Spain's wars in Europe.

By the latter part of the nineteenth century, though the silver mines had been largely exhausted, Bolivia had for years been enjoying another economic surge, this time due to the tremendous tin deposits discovered and developed by Simon Patiño and others.

The subsequent descent of Bolivia into hemispheric oblivion is deeply disturbing. Major factors contributing to its fall include exploitation of labor, and subsequent labour unrest; overprotection of industries; market collapse; war, (WW1, WW2, Korea, Chaco and many coups); the rise of a Socialist government that nationalized the tin industry in 1952; and control of that key sector by international interests. The US government, in fact, was instrumental in helping other powerful international players to destabilize the Leftist regime in question; that regime was eventually replaced by a right-wing junta.

With the help of their US and Nazi advisors, the new caudillos, or strongmen, retained absolute power. It is not surprising that Che Guevara should have emerged into this mire of violently polarized interests, seeking to bring about revolution. His Bolivian stroll in the selva was to bring his undoing.

Looking once again at Bolivia's economic picture will—no doubt surprisingly to some—return us to the critical importanceof mining in an economy which has depended so heavily upon the benefits derived from that single pursuit throughout its history. That is because, in response to the increasing volatility of the metal markets, Bolivia was forced, at a very late stage, to develop non-mineral products to diversify its economic base.

Sadly, the resource that would eventually eclipse in value all other attempts at diversification was the coca plant. What made it valuable was its use in the production of cocaine.

Hazards of Mining

While cocaine may be one of today's major "distractions" for those seeking escape, in the nineteenth century the main distraction was alcohol. Not surprisingly, the dangerous and gruelling lives led by miners drove many to drink. And their drinking was legendary. With every mining district came the obligatory hotels, bars, and houses of joy. Liquor regulations were few. The rise of the Women's Christian Temperance Union in the late 1800s marked the degree to which alcohol abuse had grown in the US. The National Prohibition Act of 1920 (also known as the Volstead Act) was the culmination of their efforts.

When one looks at the working conditions of early miners, one sees reasons for wanting an alcoholic escape. The numbing effect of twelve-hour days spent enveloped in the total darkness of wet, cold man-made caverns, hand-drilling and shoveling is beyond modern comprehension.

Of course, the dangers associated with mining abounded. There was the ever-present threat of accidental dynamite blasts; boiler explosions; rock, snow, and mud slides; fire, flood and coal damp; injury and disease. Away from the mines, a gunfight resulting from a claim dispute could cut a man down, as could a swig of poisoned whiskey. The death toll of miners was alarming.

The Major Cities-Legacies of Mining

Just as many men wrested their personal fortunes from the mines in days past, so do many urban centres owe their existence to mining booms. San Francisco, Seattle, Portland, Denver, Salt Lake City, Tucson, Melbourne, and Johannesburg blossomed as the result of mining wealth. Interestingly, much of the money generated from the mines of these cities went elsewhere—to London, Paris, Berlin or New York. Frank Lloyd Wright's non-rectilinear Guggenheim Museum in New York is a testament to this truth. The fact is that the genteel elegance of turn-of-the-twentieth-century capitals was underwritten with mining money from South Africa, the Americas or Australia.

At the same time, the Klondike Gold Rush of 1897-98 created explosive interest in Canada's north and in Alaska. The subsequent mineral-related expansion in the Alaska Panhandle, Nome and Kennecott, Alaska—as well as the rise of fishing and logging interests in the region—furthered the growth of those areas as well. The ensuing US boundary dispute with Canada underscored how valuable relatively unexplored regions were in the public mindset, and all because of mining.

Unearthing Aerial Tramways and the Path of Byron Riblet

This 'Remembrance of Trams Past' was born of an innocent 20-minute trip to the public library, the objective of which was to gather information on aerial trams in British Columbia. The lack of information yielded by that search turned the objective into a 12-year quest. Never in the author's experience had an historical question not yielded an answer—even if it took poring over multiple volumes, nor the help of reference librarians, in order to arrive at that answer.

The research for this work consequently migrated toward more local histories, tracts on mining, encyclopedias, and larger libraries. The preponderance of historic regional mining districts, one's separation from the vitality of their day-to-day operations by the passage of time, and the passing of the subject well beyond living memory all added to the complexity of the project.

Frustration, confusion, and exasperation set in. Never had so much text been scoured only to yield the words "...had an aerial tram...." Contemporary chroniclers had seemed to nod in acceptance of the existence of aerial trams, but did not expound upon the subject. Details were non-existent, let alone operating records, blueprints, or meaningful oral records. Despite all this, there were to be found a few useful records in local histories, and in specialized period journals.

The mission also made for strange encounters with those near and dear, such as when this author approached his wife to ask in which defunct mining camp she would like to vacation: "Honey, which tailings pile would you like to suntan on this summer?" This drew the natural response of, "I'd rather take a luxury cruise down the Love Canal."

Compounding the challenge was the periodic exhibition of official indifference during the course of the search. Acknowledged sources of information were consulted and either studiously ignored requests for said information, or politely dismissed them. On the other hand, there have been unanticipated rewards—and perils. This study has taken many years, and has involved encounters with truly helpful people…but also with bears, cougars, and hordes of mosquitoes.

The author also has been in close confines with wild creatures of a different sort—at times it has taken the resolve of a Special Forces soldier to withstand the withering glares and razor sharp minds of reference librarians who itemize, detail and disseminate citations faster than a Google server. (Undoubtedly, these are the same mild mannered, middle age ladies who blithely bid six no-trumps at bridge.)

Ultimately, research on Riblet and his business led to exploration of events and persons connected with a variety of distant locations and cultures. The ebb and flow of world history resonated throughout this project. Deeds both high-minded and ignoble were encountered; so, too, was tragedy, humor, and travail.

In the course of research, narratives meandered off into unfamiliar territory. Sometimes it seemed that only the constants of earth, time, a few written records, and the physical remains of old mines existed. Yet, surprisingly, repeatedly, key bits of information would suddenly come to light and further connect the pieces of the puzzle.

Byron Riblet was the product of a different age. His mores and outlook have long since passed, and indeed he felt that passing, particularly during the Depression, when his life's work seemed to be crumbling. But prior to this, he represented the positivist, progressive mentality of the turn-of-the-century West. The pre-presidential Herbert Hoover, who made his fortune as a mining engineer rigging the markets, also fell into this category; the same may be said for Teddy Roosevelt, who was to become the force behind construction of the Panama Canal and the expansion of the US Navy; and Henry Kaiser, builder of roads, dams and ships throughout the West. The wide-open West presented a *tabula rasa* on which these men and others could author their own remarkable stories. Its freshness must have been liberating.

Byron Riblet oversaw the transition of tramways from their infancy to industry dominance. Many of his mining installations have never been improved upon. It was he who built the longest systems in North America. The Riblet Tramway Company later capitalized on that experience to become market leader in skilifts.

However, mining-related aerial tramways became largely obsolete by the 1950's and they were subsequently forgotten. Their engineers and operators died out and left a patchy record behind. New mining technology superseded cableways: this included open-pit mines, heavy-haul trucks, and conveyor belts. A new generation of engineers took over, and tramways were put to use in other applications. They were used, for example, at Alcan Aluminum's Kemano (British Columbia) project to move workers, and to access remote microwave sites for the BC Telephone and BC Hydro Companies.

The tramway industry was revolutionized with the emergence of the new ski industry. New North American companies came into the business, while European firms also built tourist links. Newcomer companies like Garaventa and Doppelmayr have produced new technical developments, hence extending tramway history to the present day, and ensuring its survival beyond the current period. Yet these latecomers are only part of a lengthy and local history—a history that we will now start to examine in greater depth.

Background History,
Andrew Hallidie and the Hall Mine

1. Background History

"Toad's father may have been an admirable animal, as Badger says; but he most certainly made his pile in copper mines..."
—Peter Green in *Behind the Wildwood*

The histories of mining and technology are completely connected. For over four hundred years developments to improve mining production also spurred the use of new machinery. For example, the first railways were used as a means of easing the strain of wheelbarrow loads in the mines of Bohemia in the Late Middle Ages.

The first citation of an aerial tramway is as an illustration in a book from 1411 by Hartleib. The first documented useful aerial tramway in history is one built by Adam Wiebe and was put to use in Danzig in 1644. That tramway consisted of a rush basket conveyed overhead by ropes and was used to build town fortifications. [1] Undated Japanese woodcuts show the transfer of goods and people in mountainous country by suspended ropes; possibly making the Orient as the originator of tramways. Across the Pacific in South America, the Incas travelled on suspension bridges made from woven grass.[2] Thus the prehistory of passenger and freight tramways is obscure and unclear.

In Restoration England, Thomas Savery devised a water pump for Newcastle area coal mines. It was the first proto-steam engine; one that resembled more a pressure cooker than moving engine. Thomas Newcomen improved on Captain Savery's idea with his semi-rotative beam engine for coal mines. He had engines working by Queen Anne's accession in 1701 with accurate records of one at Dudley Castle in 1712. Newcomen used the first pistons and cylinders. Of course, James Watt continued the engineering refinements of the beam engine a half century later with the development of the separate condensor in 1765 (and, later, the rotating engine). Mining and industry flourished as a result of these mechanical improvements.

All throughout the nineteenth century Newcomen and Watt engines gave excellent service (some lasting in use for over one hundred and thirty years) and they were further refined with the Cornish cycle. Winding engines conveyed the men and the material up and down the shaft safely.

> Rotative beam engines used to hoist the ore were called 'fire whims' to distinguish them from the old horse whims used previously. They were also used to drive batteries of stamps for crushing the ore or for pumping lesser quantities of water. Some stamp

engines and whims also drove pumps and had a second beam projecting from the rear of the house for this purpose. Another quaint device which took its drive from the rotative engine was a 'man engine' employed in some of the deep mines to move men up and down the shaft. In this a vertical timber rod and small platforms and handles attached at fixed intervals was driven up and down by a crank at steady speed, the platforms coming level with fixed platforms or 'sollars' in the shaft at the top and bottom of each stroke. By stepping on and off during the momentary dwell between strokes men could travel up or down.[3]

Blower engines removed the toxic fumes from the powder blasts. And, as always, pump engines laboured to keep the waters out.

Richard Trevethick, a mining engineer from Cornwall, tinkered with road engines and the first steam locomotive in 1804. He later worked as a mining engineer in South America. While George Stephenson, a semi-literate mining millwright from Newcastle took the idea of the railway engine further. He had grown up watching the coal tubs being pulled on their railways around the collieries by horses and with manila rope hauled by winding engines. Thus, he too began devising passenger railway engines. His "Locomotion #1" hauled the first passengers on the Stockon and Darlington Railway in 1825.[4] Large suspension bridges were being engineered at this time too. The first half of the nineteenth century saw the rise of the field of civil engineering. Thomas Telford built his Menai Straits bridge in Northern Wales in 1825. As iron cables large enough to support bridge spans were not available, wrought iron links were used on the Menai bridge. The links resembled a large bicycle chain and upheld the bridge deck.

Isambard Brunel's career started slowly first assisting his father, then erecting a Watt pumping engine for a drainage scheme in Essex. He then built, what some consider the finest railway in England, the Great Western Railway. Following that success, Isambard Brunel built some of the first sea going steam ships—the *Great Western*, and the iron hulled *Great Britain*. For the purposes of this paper Brunel's activities indicate the state of engineering in the mid nineteenth century. Many of his projects were ahead of their time: the Saltash Bridge, the elegant Clifton Bridge at Bristol, and the massive ship–the Great Eastern. The Clifton Bridge (completed in 1864 after Brunel's early death) uses chains to support the bridge deck. The steamship *Great Eastern* (1858) was so large, just short of 700 feet in length and displacing 32,000 tons, its launch was fraught with difficulty. The hull was

A medieval Japanese tramway.

not exceeded in size until the end of the nineteenth century. Launching such a leviathan required many winches, chains, and hydraulic jacks. The most famous photograph of Isambard Brunel (in the age of Henry Fox-Talbot) is of the engineer confidently framed with a backdrop of stout launching chains. Iron and steel cable were products of the future. Despite her size, the *Great Eastern* had a tarnished career. Her main claim to fame was acting as a cable ship when she laid the first woven copper, iron, tar and jute telegraph cables across the Atlantic in 1865-66. The cable failed on the first attempt, though a second attempt with another cable led to the successful completion and use of both cables from Newfoundland to Ireland.

Improvements were also tried to the aerial tramway. In 1825 Dr. Purkinje of Vienna used rope in an early funicular railway. While a mine engineer named Albert—in the Hartz Mountains of Germany—first used steel wire for haulage; to this day a type of cable has a weave called Albert Lay.[5] In England, a man named Robinson patented the monocable rope system in 1856. Then another man called Hodgson used this patent to build what is thought to be the first usable tramway in 1857. Hodgson laboured over the years, and together with the inventor Roe, formed the basis for the successful monocable designs of Ropeways and British Ropeways Engineering, which exported systems around the globe for decades in the twentieth century.

In America, Andrew Hallidie was a contemporary of Hodgson and Hallidie built some of the first workable aerial tramways in Western North America and dominated the North American market for around 25 years. Andrew Hallidie will be discussed at length in the next chapter. After Hallidie's designs became obsolete around 1900, the firms of Trenton Iron Works, Leschen and Riblet competed in the mining tramway field.

More developments relevant to mining continued in the nineteenth century—these include the advent of nitro-glycerine replacing black powder as a blasting agent. Later, by 1866, Alfred Nobel's dynamite negated the negative effects of unstable nitro-glycerine. One of its first large scale uses was driving the Mont Cenis Pass railway tunnel in Europe. While San Francisco experienced at blast in 1866 when someone shipped the liquid nitro-glycerine by Wells Fargo Express and it exploded in the city causing death and damage.[6] There were many other tragedies with the liquid blasting agent in the Victorian era.

Blasts of a different sort were reduced with the advent of Safety Lamps; invented by Humphrey Davy and George Stephenson. These lamps reduce the risk of danger from explosions when the open flame of a candle came into contact with methane gas that is so common in coal seams. The coal itself was used to power industry and to smelt steel.

That steel became available, for the most part, as Henry Bessemer's blast furnace revolutionised the iron and steel industry in 1872 and made large-scale production of steel possible. That steel would revolutionize all industries from smelting, to railways, to bridgebuilding, to ship building, to the proliferation of steel cable. In turn, the development of these industries would spur the demand for all metals.

Vast improvements in mining and tunnelling came with the steam drill. Drilling into rock became easier than 'hand jacking'—the slow hammer and chisel method of drilling blast holes in rock. An American J. Couch patented a spear like steam drill in 1845. Then J. Fowle, a mechanic who worked for Couch improved upon Couch's design. Subsequently, Burleigh bought the Fowle patent rights and designed what is thought to be the first workable rock drill. Compressed air drills later replaced steam drills; (they were troublesome and not as reliable due the fact that steam had to be piped underground to the working face. In the process the steam condensed.) In Europe, Alfred Brandt designed a hydraulic powered drill and it was put to use on the Simplon Railway tunnel under the Alps.

Railway tunnels were the testing ground for new rock drills. In North America, the tunnel of note was the Hoosac Tunnel in Massachusetts. That bore was an ongoing trauma which took 24 years to complete, cost 195 lives, and bankrupted the Commonwealth of Massachusetts. The mountain's solid granite successfully resisted the designs and technology of men for two dozen years until 1874. Only the advent of dynamite and powerful Burleigh drills allowed for the success of that passage. The Hoosac Tunnel was the longest in the world when built and was only exceeded in length thirty years later by the five mile Spiral Tunnels under the Canadian Rockies.

The 1870's were a period of business contraction. Overbuilding of railways inflated the market to where the bubble must burst. This happened in 1873 with the Credit Mobilier scandal—caused by the Union Pacific Railroad stock fraud—and which sent other railways into bankruptcy. One foreclosed line was the uncompleted Northern Pacific Railroad then striving towards Portland and Seattle—it suffered and went bankrupt as well. It was not finished until 1883. The ensuing economic depression chilled investment and caused mass unemployment. Not surprisingly, there were violent railroad strikes and industrial disruptions. One ray of hope at the time was the prosperity in the gold and silver mines of Nevada, Utah, and Colorado. Out west the mines boomed and the legends of Cripple Creek, Leadville, Creede, and Silverton were secured.[7] Similarly, at the mining camps, Hallidie tramways were built as mining flourished in mountainous areas.

When the Northern Pacific Railroad was finished across the northern US States of Dakotas, Montana, Idaho and Washington, hard rock mining expanded into Idaho. Subsequently, the Coeur d'Alene mining boom of the 1880's spilled over into Canada, prospectors combed the then unexplored region looking for metal bearing quartz. The first large mineral deposit find in the Kootenay Region of BC was at Riondel and Nelson. American prospectors, flush with success from the Coeur d'Alene mines, scouted the southern end of BC and located the large silver deposit at Toad Mountain in 1886. Not long after the deposits at Rossland were located. Soon the famed War Eagle and LeRoi mines began production there securing the Kootenay as a valuable mining territory in the world. In the end, these mines resulted in the building of many Hallidie and Riblet mining tramways.

The other spectacular late nineteenth century mining boom was the Klondike Gold Rush. The year 1898 brought the Klondike Gold rush. Tens of thousands of mainly American gold seekers left their homes to try their luck in a dangerous, long distance venture. Those numbers are a testament to the shaky state of the US and world economy following the "Silver Panic" of 1893. The punters had to first travel to Seattle, then by old, leaky chartered ships to Skagway. All the time avoiding the merchants and tricksters swindling the last of these unfortunate's money. Then came a horrendous series of tests and travails just to get to the goldfields of Dawson—the most famous being the Chilkoot Pass. This was a steep incline over which the sourdoughs had to climb and carry heavy loads of supplies. As a means of improvement, three aerial tramways were built to haul supplies up over the Pass. While not remarkable in mining circles, the Chilkoot Tramways received much press.[8] The tramways only operated for one year before they were made obsolete by a narrow gauge railway.

John Roebling—Wire Rope Hero

Another towering figure in wire rope history was John. A Roebling. Roebling was a German immigrant who came to the US in the 1830's he established an important wire rope business. His Roebling and Sons Wire Factory became the most important in the country. Roebling also made a name for himself by building some of the first large suspension bridges in the Eastern US.

Imbued with the possibilities and elegance of the suspension bridge from an early bridge in his beautiful medieval German city of Bamberg, Roebling set forth to do the same in the New World. Educated by Hegel in Berlin, where he studied engineering at the Polytechnic, Roebling had a sense of destiny. He set the pattern with a bridge at Cinncinati and together with the engineer James Eads (who built the first bridges at Niagara and over the Mississippi) incrementally increased bridge engineering capabilities. As many of the expanding cities of the US were situated on large rivers, they needed bridges. Roebling would cap the trend as he was the visionary behind the Brooklyn Bridge.

The Brooklyn Bridge was necessary to help expand the burgeoning population in lower Manhatten over into the available land of Brooklyn. Ferries carried the people, though a fixed link would improve the quality of life on both banks. Started in the 1869, the bridge pushed the capabilities of engineering as caissons were needed to sink the piers into the mud under the East River. Workers suffered horribly with the unknown phenomenon of the bends, as did Roebling's son—who would complete the bridge after John A. Roebling died prematurely. The two soaring gothic stone towers took years to finish. Yet it was with the cables that Roebling's genius came to light—each wire was individually spooled across the river. Wires were bundled into cables, four in total on the bridge. The spinning of cables on bridges became standard practice in civil engineering for almost a century. The bridge opened in 1883 and is a landmark to this day in New York. Fittingly, a cable car was installed to ferry passengers across the bridge from Brooklyn to Manhattan.

As stated, Roebling established a large wire factory in New Jersey by 1848 and he prospered with the expansion of America business by supplying cable to industry—including aerial tram projects. America had a large demand for telegraph wire and later barbed wire. Such was the factory's notoriety and importance that in the First World War German saboteurs torched the wire plant in 1915.[9] They also destroyed the munitions loading dock at Black Tom on the New

Jersey shore. In the process of the resulting explosion, pieces were torn from the Statue of Liberty.[10] It turns out there are other ground zeros within sight of the lower West Side.[11]

The Roebling Company would extract its revenge on the Germans in kind. When America joined the war in 1917, the U-Boat menace was reaching a crescendo and causing the British Admiralty much concern.

A revolutionary solution was to create a wire rope net and minefield stretching from the Orkney Islands to Norway. While the British were skeptical of such a large undertaking, the US Navy boldly stepped in and built it.

> **The [naval] mine field would measure some 230 miles long by 25 average width, consist of 70,000 mines in systems, each consisting of one or more mines near the surface, other mines deeper, and yet more deeper still so as to bar or imperil the passage of any vessel either surfaced or submerged.**[12]

Secret paths were left through the Northern Mine Barrage to allow for friendly traffic. In order for the system to work, thousands of miles of wire rope needed to be manufactured to secure the nets and mines. Roebling's factory alone turned out 27 million feet of wire rope for the project.[13]

The ultimate effects of this barrier, and aggressive Royal and Allied Navy patrols in their steam powered submarines and other ships, was success.[14] The German High Seas Fleet rarely ventured from port after 1916, nor were commercial ships bringing in cargos. As a result of a lack of imported food and fuel, the populations of the Central Powers in Europe starved between 1917–1919. One victim was the infinitely capable mathematician Georg Cantor. Additionally, it has been argued that the resultant malnutrition in the children lead to weakened mental capacities which some suspect, allowed for the population to be led astray in the 1930's.

Terrorists seemed to be attracted to the Roebling name. In 1953 Cominco—the large mining operator in the Kootenays—built a power line from its hydro dams to a lead-zinc mine at Kimberley, BC. This involved a two-mile powerline crossing of Kootenay Lake near Balfour. It was one of the largest over water spans in the world. So large was the span that the conductor wires were made of steel by Roebling, and not of the usual copper as the wires needed strength to hold up their own weight. The cables were transferred from a high bluff on the West shore over the lake to a 375 foot tower on the East shore. The tower was built to shorten the span as it was pushing the limits of engineering at the time.

Early one morning in 1962 an angry, schismatic sect of Doukhobors placed dynamite on the tower supports and exploded it. As a result, the tower collapsed and the electric lines fell into the lake stopping the flow of power. The Canadian Army was called out and other infrastructure points were guarded. The power line was rebuilt. The distrust between the Doukhobors and government remained. The Doukhobors were angry as they were settlers originally from Russia and were promised religious freedom. They lived communally, did not believe in government, and were prone to bouts of religious fervour. The feud with the government had been going on for years, starting in Weyburn Saskatchewan in 1903 and then later moving to BC. Included in the dispute were issues of land control, oaths of allegiance, education, arsons, bombings, arrests and internment camps.[15] Eventually, the conflict subsided with the arrest of hotheaded activists.

Insurrections and sabotage in the West were neither limited to warring nations nor to local religious sects—there was also industrial action. The rugged conditions of mining in the West, combined with the insular nature of company towns that allowed those companies to act with little heed for others, led to major conflict.

> The violence thrown against the metal miners of the West as they fought for a living wage and the right to organize during these years of gunboat diplomacy is almost without precedent in American history. The sagebrush plains and alkali deserts of Utah, the little one street mining camps high above the timber line in the Colorado Rockies, the gulches and canyons of northern Idaho, all heard the tramp of invading, strike-breaking troops. The regular army and the National Guards of the various states, particularity in Colorado and Idaho, were called out at least a dozen times in a dozen years. Either federal or state troops were sent against the miners in the bloody strikes of Coeur d'Alene in 1892, of Leadville in 1896, at Salt Lake City and Coeur d'Alene again in 1899, at Telluride in 1901 and 1903 in Idaho Springs and Cripple Creek in 1903.
>
> Thousands of miners were placed in barbed-wire camps, as at Coeur d'Alene in northern Idaho in 1892 and 1899, where they were held for months without trial or charges being placed against them Other thousands were herded from their homes at the point of a bayonet, as at Cripple Creek and Telluride in Colorado in 1903, loaded onto freight cars as if they were cattle, and deported by the

military without trial or charge except that they were union men. In the great 1903-04 strike for the eight-hour day in Colorado 42 men were killed, 112 wounded, 1345 arrested and imprisoned in bullpens, or military concentration camps, and 773 deported from the state.[16]

Despite such high-handed techniques, the industrial insurrection was not curtailed. The governor of Idaho, Frank Steunenberg, was assassinated with dynamite affixed to his garden gate by an alienated miner who was aggrieved by the strikes in the Coeur d'Alene.[17]

Leading these strikes were some charismatic leaders. Three important union leaders were William Haywood, Samuel Gompers, and Eugene Debs.[18] Each fronted a different union but upheld a socialist attitude. The division between craft and unskilled labour, wider politics and the narrow focus of the workplace split the socialist platform and weakened the movement. Gains were made with the legislation of the eight-hour work day, free speech, and later worker's compensation laws.

The strikes continued with labour unrest in Colorado reoccurring after 1912. A young Canadian labour lawyer named William Lyon Mackenzie-King was hired by Rockefeller to "negotiate" and establish harmonious relations in his mines. The Rockefeller mines at Ludlow, Colorado were the scenes of a bloody strike in 1913 where women and children were gunned down by the US Army.

Radicalized workers stepped in to fight against the entrenched bosses and the armies of the state. Two famous workers were Ginger Goodwin and Joe Hill. Both worked in the mining industry and were involved with strikes in the US and Canada. In the course of events both were killed by actions of the state.[19] Ultimately, labour peace with the miners and industrial workers was first delayed by the First World War, when martial law paved the way for the arrest of labour leaders, then obtained by labour shortages in that war and the eventual post war prosperity in North America in the 1920's. All the same, at the time the strikes were bitter and divisive.

Bitter strikes and steel cable also caused scandal for other BC Premiers. The early BC Premier, Edward Prior, was ejected from office—for he assigned a Chilcotin Bridge contract with cable and rigging outlays to his own hardware company. That scandal marred his term in office. William Bowser promulgated a divisive coal strike on Vancouver Island in 1912; strangely, angry Kootenay miners briefly kidnapped Richard McBride during his term in office.

Thus BC Premiers complete our background discussion on wire

rope, tramways, wars and historic people; the scene is now set for the last decades of the nineteenth century and the entrance of Andrew Hallidie and Byron Riblet.

2. Andrew Hallidie

One of Scotland's unsung sons was Andrew Smith Hallidie (1836-1900). Born to an engineer in Lockerbie, he gained mechanical schooling and practical workshop experience under his father. The father held patents for the manufacture of iron wire rope.

In 1853, following the Gold Rush boom, both son and father travelled to California. There Andrew tried his hand at mining but he was unsuccessful. Out of necessity he relied on his mechanical skills and opened a blacksmith shop. Then "when he was only nineteen, he constructed a wire suspension bridge and aqueduct of 220 feet span..."over the American River.[20] Later, when he noticed the wear on a manila rope on an inclined surface tram at a mine, he suggested using iron wire rope. That rope gave much better service. Sensing the business possibilities, Andrew Hallidie went into the wire business around 1857. The success of the American River bridge led to work on many suspension bridges around the West Coast. Bridges were constructed "across the Bear, Trinity, Stanislaus and Tuolumne Rivers."[21]

Andrew Hallidie prospered with the wire business. The silver mines of Nevada required haulage rope to haul men and ore up from the depths. Hallidie specialized with a flat iron wire winding rope that was used on the vertical winding engines. This rope gave good service and was used widely: particularly in the Comstock Mines of Virginia City. He established first the Andrew Hallidie Company in the 1860's and later the California Wire Company.

With the discovery of placer gold in British Columbia in 1858, the nascent colony became a hive of activity. As the miners first worked the bars of the Fraser River, then the Quesnel Region, communication with the interior of BC became of paramount importance. A brigade of Royal Engineers was sent out from England to help develop the area. One task they were detailed to do was construct a wagon road up the Fraser River from New Westminster to the goldfields. A huge obstacle was bridging the wide Fraser River. To this end Hallidie was awarded the contract in 1863 to build a suspension bridge at Alexandra in the Fraser Canyon. Trade in BC at the time was oriented with American California due to the distance from England and the perils of 'rounding the Horn' as all cargo ships had to do at the time.

The wagon road up the Fraser Canyon into central British

Columbia was a difficult piece of engineering. The sheer rock walls of the canyon, heights above the river, and lack of labour [labourers had a habit of wandering off in search of mining riches] were but a few of the problems. The Royal Engineers persevered, cleared, graded, levelled and produced a usable turnpike. At times the road was elevated on timber cribwork clinging to canyon walls. The road was to become a central artery in that Province until the completion of the railway in 1885.

Hallidie too had problems in bridge construction. Hallidie himself wrote—in a manner reminiscent of Joseph Conrad:

> Everything of iron or steel for the bridge was prepared in San Francisco and shipped by steamer to Victoria, Vancouver Island—which at that time was a free port [no duties]—thence by another steamer to New Westminster on the Fraser River and thence by light draft steamer to Fort Yale. These latter steamers were owned by Captain Wright, who was generally called Bully Wright.
>
> The material for the bridge formed a pretty good load for the stern-wheel steamer, but everything went well until the on third day we reached Emery's Bar, about three miles below Yale. Here the stream proved too much for her. Spring lines were strung out and every device known to steamboat men tried without success—even a barrel of pitch was broached and fed into the furnace to keep up steam and a sixty-three pound bundle of wire was hung on the safety valve [lever]. The heat of the fires blistered paint and drove the passengers clear aft, but all without effect, and the captain...decided to land his cargo on the Bar.[22]

Despite the numerous attempts to have a boiler explosion and deliver the wire beyond Spuzzum against the powerful river by steamboat, they were unsuccessful. Another method had to be found.

The solution was to turn to the First Nation Sto-Lo people.[23] The aboriginals were unsung heros as they freighted, packed, guided, and traded in the pioneer era. Hallidie had cable spools placed in a purpose-built frame between two indian freight canoes, thus forming a cargo catamaran. Under the direction of Captain John—whose native name in Halkomalem was Stoles—work proceeded. With native paddlers, the heavy load was moved further up the river. Nonetheless, before the bridge site at Alexandra could be reached, the current became too strong for even the native paddlers. Again the spools were unshipped onto the bank. Ultimately, the cables were unspooled and the wire hand-carried, centipede style, to the bridge site.[24] Masons built stone pier which held up firwood towers, over which the cables were slung. Vertical iron stringers connected the parabolic suspension cables to the wooden, bridge cross members. A wooden bridge deck completed the span. It was BC's first suspension bridge.

Hallidie went on to complete other firsts. He patented a suspension bridge in 1867 and other inventions "among these was the Hallidie Ropeway a method of transporting ore and other material across mountainous districts by means of an elevated, endless travelling line, which he had invented [sic] in 1867."[26]

No history of mining and aerial tramways can omit the contribution of Andrew Hallidie. In the late 1860's Hallidie developed an wire rope aerial tram. It used a single loop of wire, much akin to a closeline, with buckets bolted to the wire with steel tabs—thus, as the wire travelled continually, the buckets moved. Nonetheless, many technical details had to be worked out starting a loaded tramway from dead stop, loading moving buckets, and wire slipping on the end bull wheels. For the latter, Hallidie patented his elegant "improved grip pulley" in 1870.[25]

Hallidie proceeded to supply the mining world with aerial trams. Robert Trennert, in his well-researched book, *Riding the High Wire: Aerial Mine Tramways of the West*, suggests the first was built at the Freiberg Mine in Eastern Nevada in 1872.[26] In quick order another tram was built for the Morning Star Mine.[27] A third tram was installed for the Vallejo Tunnel and Mine Comany in Utah by Hallidie in 1872. Trennert states that it used five-eighths inch diameter cable and was capable of moving sixteen tons of ore per hour.[28] Hallidie, and his aerial trams, became widely known.

Accessible deposits of coal were discovered at Nanaimo, BC on mid-central Vancouver Island by the Hudson Bay Company. The coal seams of Nanaimo extend under that beautiful island shelf for miles. Later, James Dunsmuir and his son Robert developed the region and shipped coal throughout Puget Sound, the Gulf of Georgia region and to the profitable market in San Francisco. As the seam at Nanaimo worked out, Dunsmuir drove a new shaft at Wellington a few miles north. Railways were built to move the coal. Indeed the first railway in BC was built to move coal at Nanaimo. However, getting the coal down to the dock with a minimum of effort was troublesome. Wellington is situated on the shelf above Departure Bay—the natural and accessible anchorage in the Nanaimo area. Consequently, part of the railway included an early incline for shipment.

Also, Hallidie was contracted to build an aerial tramway to haul coal from inland seams near Wellington down the hill to the sailing ship dock. Completed by the end of 1874, the pyramidal towers and small iron buckets meandered their way down to the harbour. The line passed over cottages and pastures while the towers competed with the remnants of the once proud and looming fir forest. Alas, the tram engineering was not up to the rigours of the field. Overloaded buckets caused the cable to sag between towers and stalled the system. In a word, this tramway did not work very well. Its use was discontinued and it was forgotten. Only a brief mention in a period newspaper and a single photograph are the only surviving record of BC's first aerial tram.

Despite the mixed success up north, Hallidie went on to build many other trams. Hallidie trams were built at mines at Sumpter, Oregon; Bodie, California; Elko, Nevada; and Banner, Black Lake, and Custer, Idaho to mention a few places.

Purportedly, one night in 1869, Andrew Hallidie saw a team of horses struggle as it pulled a street-car up cobblestoned Jackson Street attempting to ascend San Francisco's steep slope. Despite the driver's whip and acidic epithets, the horses slipped and were dragged downhill by the weight of the car. Hallidie reflected on his mining experience and then devised a plan to haul passenger street-cars with cable. He would build the first practical cable car to haul streetcars up fashionable Nob Hill.

There was much designing to do, Hallidie had to: bury a cable conduit in the street, rig a cable and pulleys, design a grip and breaking systems, and power it all with a stationery steam engine. His company and business connections underwrote the enterprise and he managed to install the world's first cable car in 1873. Cable cars are not aerial trams but 'burial' trams. Most of the appurtenances are buried under the street: the wire, cable supports, guides, angle stations, and power plants. Only the car rails and cable conduit lip are exposed to the street and mud. A half-inch gap in the cast iron cable conduit allows the car grip to clasp the cable and connect with a link to the cable car. In this way, the streetcar can move up the hill at the speed of the cable, while it can release and stop to load passengers and re-grip to continue movement. In Hallidie's day a steam powered winding engine atop the hill laboured to keep the cable moving at around six miles per hour.

The benefits of the cable car system over the horse drawn manner were obvious. More lines were built by other engineers on San Francisco streets. Soon street car companies were established in other hilly cities such a Portland, Los Angeles, San Diego, Seattle and

Tacoma. The technology proved so reliable that cable car operations were set in operation in Chicago, New York, Pittsburg, Kansas City, Denver, London, Lisbon, Melbourne, Dunedin and Sydney as well. Supplying the necessary sheaves and cables became a steady business. The drive cables had to be replaced every year, and sometimes more often if the cable had tight corners or muddy applications. Thus numerous suppliers went into the business in addition to Hallidie: Leschen, Broderick and Bascom—both of St. Louis—Roebling and Sons of New Jersey. Each company extended their street cable car business into the mining aerial tram sphere.

Hallidie's and San Francisco's first cable car on Clay Street: It was built up on Nob Hill in 1872. Note the forward grip-car with operator. After this more cable car lines were built.

Only the advent of reliable DC electric motors in the late 1890's, and early 1900's provided an easier technology which finished the reign of the street cable car. Of course, cable cars survive as a curiosity to this day in San Francisco.[29]

3. The Silver King Mine, B.C.

The last quarter of the nineteenth century was a transition period in mining in Western North America. Previously the miners scraped and hand shovelled for gold in the placer streams. Such waning placer undertakings were the impetus for exploration through the 1880's. A party of twelve men—led by William Hall—left Fort Colville in Washington State and prospected around the Kootenay in BC looking for rumouredplacer deposits.

Camping one night near the southern end of Kootenay Lake, Hall's pack horses escaped their tethers and wandered off looking for forage. In the process of corralling the horses the next morning, the men noticed rich outcroppings of silver. This mine became the famed Silver King and it underwrote and became the being for the City of Nelson. With the claims staked and a rich showing on the silver vein atop Toad Mountain, adits were driven there by 1889. It must be said that miners were also at work on the Bluebell Mine at Riondel, in 1886, making it the first mine in the Kootenay District.

By 1890 Nelson had cabins and steamboat links to Idaho and subsequently became a booming mining camp. This group of Toad Mountain claims, above Nelson, were assembled into the Silver King Mines and production began. However, there were logistical problems with the Silver King mines (or sometimes Hall Mines.) A pack trail was in use, and it was upgraded to a full road to allow for the use of wagons. The Silver King mine was situated at some 4800 feet in elevation on Toad Mountain and the transport of supplies in and ore out posed a large problem.[30] As usual, costly pack horses conveyed the freight up the seven mile service trail.[31] In the first years of the 1890's the Silver King was shipping its ore to smelters in Montana. It was quite an involved process in transporting the heavy ore from the mountain to the dock, across the lake, to a river system, another dock and eventually a railroad. To reduce shipping costs technology was employed. Thus in 1895 a four and a half mile Hallidie aerialtram was constructed from the Silver King mine down to Nelson.[32] It was a monocable design, with small, square, hundred pound capacity ore buckets. It is claimed that it was the longest Hallidie cableway in the world when it was built. Nonetheless, as with the tram in Nanaimo, it was far from perfect.

The long length of the tram, complete with huge, unsupported spans across ravines, led to sagging of the cable and problems with grip on the bull wheels. There were also problems with the cable slipping off the tower sheaves.[33] With these drawbacks, the tram was eventually rebuilt in two sections. The strain of the long descent contributed to the problems so a mid-station was installed. In effect, splitting the tram into two separate trams with the mid-point being a reload station. In this way the strain on the cable and towers was reduced and the tram worked satisfactorily afterwards.

Hallidie Tramway at work on the Silver King Mine.

Mid- station of the Silver King Tramway. It was the largest tramway in the world when it was finished and it hauled silver ore until 1902. BCARS C00581

Hallidie's California Wire Company supplied the 50,000 feet of cable needed for the tramway and it was shipped from San Francisco. By 1895, there was rail communication with Nelson as the Nelson and Fort Sheppard Railroad entered that City from the US some two years earlier. In addition to the cable, some 126 support towers were needed to elevate the cable in the swath cut through the trees.[34] A metal bracket was woven into the cable at intervals from which each of the buckets could be affixed.

Ore was carried in the cubic buckets that numbered a staggering 900 in total. These delivered about ten tons per hour of ore to the bottom ore bin. Hallidie's patented grip bull wheels were used at either end. With the monocable design, the cable was always in motion, and, as the buckets were permanently attached to the monocable by bolted metal brackets, the buckets had to be loaded and unloaded while in motion.

There is one spectacular photo of the tram, taken presumably when it was first built, showing the tram high above Nelson and overlooking the West Arm of Kootenay Lake. The tram is indeed a rough affair, made of rough-hewn local timber, while the operators look like they are made from even coarser material. Whatever the shortcomings of the operation and miners, the tram worked, hauled the ore and made the mine wonderful profits.

There is one grisly story about the Silver King tram involving a traveller hitching a ride up the mountain on one of the buckets. Somehow, in the process of moving up the mountain, the traveller was entangled between the tower and a bucket as it tried to pass. This occurred in mid-journey, well away from the stations. Due to the blockage, the tram stopped. The operators were obviously unaware of the traveler. To fix the problem they decided to activate the cable with a chain hoist from the terminal and clear the blockage. The unlucky, jammed journeymaker was then distended bodily by the mechanical action. By the time it was discovered, it was too late.

A smelter was built in 1895 in Nelson itself and the retort was 'blown in' or started in early 1896. Yet it was strangely obsolete, with the completion of A. Heinze's larger smelting complex down the river at nearby Trail in 1898. Thus smelting in Nelson stopped. The Silver King mine had a large output until is shut down in 1902. After that it operated intermittently under lessees, though its output was low. Nelson, by this time, became the *entrepôt* centre for the entire region.

The Mines of Rossland

In addition to the Silver King mine at Nelson, the mines of nearby Rossland were important. Rossland is near the intersection of the US Boundary and the Columbia River nearly sixty miles southwest of Nelson. Despite a period of speculation, investors believed in the region and capitalized the Rossland mines. Two important mines were the War Eagle and the Le Roi which developed workings on Red Mountain. In the process, the Le Roi mine was developed and it installed a huge concentrator and industrialized minehead. Between the two operations ran a Hallidie tram. Today, a mock-up replica of the tram stands at the Rossland Museum. Included in these businesses was the building of the smelter on the river at Trail and railway to connect all sites including the mines at Nelson.

It seems that there was a scandal and that the Le Roi Mine was at the centre of it. The Le Roi Mine was purchased in 1897 by the British America Corporation, a London based consortium headed by one James Whitaker Wright. He subsequently issued new stock in a new company for that mine. However, in the next two years, the mine operated at a loss. It took several years for the financial position to be revealed which happened by early 1901.[35] Whitaker Wright was overextended and went bankrupt; he was convicted of fraud and, spectacularly, took cyanide in the courtroom.

Rossland is also famous for its skiers too, having produced several Olympic winners over the years. Skiing and Rossland have a long history—indeed one of the first ski chairlifts in BC was at Rossland.[36] Consequently, Nelson and Rossland formed a foundation for the Kootenay Region with the establishment of their mines and the building of early tramways, things that provide a background for the entrance of Byron Riblet.

4. Pullman and Tekoa, Washington

Preamble over, now is the time to introduce our principle character. Byron Riblet was born in Iowa. He grew up on a farm and was schooled locally. He attended the University of Minnesota where he took a degree in Civil Engineering. He worked his way west with the expanding railroads and by 1885 found himself in the US Pacific Northwest.

One of Byron Riblet's first jobs was to act as civil engineer for a spur line for the Northern Pacific Railroad. As an inducement to build track, the railroads were given land in the form of grants by the government. The Northern Pacific received some 39 million acres. It then sold parts of the land to settlers and timber barons like Weyerhaeuser. After the railroads finished their transcontinental line, they would proceed to expand their network with extra railroads to act as feeders for the mainlines. This is how a short spur into the wheat country of Washington State's Palouse territory was built. It ran from Spokane to Pullman.

At the time, the competition between the Northern Pacific Railroad and the Oregon Railroad and Navigation Company was legendary. The O R and N Company had earlier trumped the NP Railroad with a line to Colfax in 1884. To overcome this loss, the NP incorporated the Eastern Washington Railway with an eye to build a line from Spokane to the Snake River. During the legal process, officers were elected, stock subscribers found and the name of the railway duly changed to the Spokane and Palouse Railway.[37]

Construction commenced in the spring of 1886 on a route surveyed from the mainline at Marshall, through Rosalia, Garfield and Pullman. The route was 'covered in men and horses' as the crew scraped and graded the right of way through the pleasant rolling hills. Byron Riblet worked as an locator and bridge engineer for the line.

Teri Hein—in her description of life in the Palouse and its later experiences in *Atomic Farmgirl*—explains the nascent railway and its effect to the region.

> When they [her great grandparents] arrived in Washington Territory, the railroad had run a track to a platform named Regis, which later became the town of Fairfield, ten miles to the northeast of what is now our farm. Having a railroad platform was essential for farmers who wanted to send their wheat into the world of consumers.
>
> In those days the railroad companies were inventing towns right and left, hoping to lure pioneers to these barely charted lands, where they would start farms and use the railroads to ship their wheat. Lonely platforms in the middle of prairies were given impressive names such as Ritzville and Grandview.... Or platforms were given the names of railroad executives or their friends—for example, Cheney and Pullman.[38]

Such was the process of development in the West.

The NP Railway and the O.R. and N Railway continued their feud with court injunctions, sabotage and gunfights on the grade. In the end they managed to divide up the booming wheat trade. Byron Riblet finished the spur to Pullman in 1887.

Tekoa and the Coeur d'Alene Railways

With the booming mines in Idaho, again the Northern Pacific Railroad and the Oregon Railway and Navigation Company also wanted mineral traffic. Railways were built from Spokane and Montana to the Idaho Panhandle district. The O.R. and N. Railroad proceeded with designs to build a line from Tekoa, on its Palouse line near the eastern border of Washington State, to the mining district. This was to become the Washington and Idaho Railroad. One time consuming difficulty was the process of acceptance of the right of way through the Coeur d'Alene Indian Reservation.[39]

The Washington and Idaho Railroad would travel from Tekoa along Hangman Creek and over into the St. Joe River basin. It then skirted the Southern End of Coeur d'Alene Lake and past the Cataldo Mission to Wallace. Byron Riblet was the divisional engineer in charge of construction from 1888 to 1889. Problems continued with the rival Northern Pacific, who harrassed the competing company's labourers by rolling large rocks at them while they attempted to work.

The duelling railroads were jammed together in the narrow canyons that formed the mining communities in the Bitterroot Mountains. The town of Burke had so little space that the line went up the main street in town and the hotel had a section of the building removed to allow the trains to pass. In such cases the railways signed joint operating agreements where there was only room for one right of way.

The coming of the railroads would neither bring peace nor prosperity to Northern Idaho. In the course of events, metal prices shrank which affected profits. Also, railway overbuilding in the mountainous country pushed the Northern Pacific into bankruptcy. The financial depression of 1893 was a severe one.[40] Reduced profits forced the mine owners to pass the shortfall onto the labour force with a wage rollback. The ensuing strike became history with guns, violence, arrests and death. By this time, though, Byron Riblet was in general engineering practice building dams, power systems and street railways elsewhere.

5. Streetcars in Spokane

The completion of the Northern Pacific Railroad across Washington by the last quarter of the nineteenth century provided the impetus for the growth of new towns. Spokane had previously been a convergence point and the site of a Hudson's Bay Fort and became a railroad center. In time three separate railways would bring traffic through the town. Thus Spokane had and has a bright future. Late nineteenth century land development mirrored the expansion of street railway networks, and this was true for towns throughout the West. Land was assembled, parks and features installed, the streetcars established and then lots sold to finance the venture.[41] And so it was in Spokane—the Spokane Street Railway was started in 1886 with the use of horse pulled streetcar. Lines were extended to the elegant Browne's Addition and to Coeur d'Alene Park.

The success of the first projects heralded a wave of other street railway companies. Soon there were three separate lines including a cable car. Of course, there were altercations over competing rights of way and fist fights between construction crews. In turn, the legal wrangling between rivals started, and left to preside over the trouble was the city council.

One of these companies was the Ross Park Street Railway, with a railway along Trent Street, east over the Spokane River continuing north on Hamilton Street to Ross Park. Byron Riblet was hired as engineer for this extension helping to layout track and design electrical equipment. The line opened in 1889. Also, carriers were needed to move people, as a result, Byron Riblet tried his hand at design.

> **The eighteen new trolley cars which were ordered from J. B. Brill Company of Philadelphia were designed by B.C. Riblet of Spokane. They were to be two truck cars, with fifteen horsepower motor on each truck, giving the cars a total of thirty horsepower. They were to be open sided on the ends and enclosed at the center of the body, being a design similar to the cable cars then in operation in many cities, including Spokane.[42]**

The Spokane Street Railway bought out the Ross Park Line in 1892. It continued its expansion of service with more lines, and the construction of a famous bridge across the Spokane River on Monroe Street, just West of Spokane Falls. In the process Spokane grew to be the modern center it is today.

6. Washington Water and Power Company, Spokane

With the establishment of Spokane in the 1880's, a city was begun on the frontier. This burgeoning metropolis included electricity service for lighting. At first a small dynamo was put to work for that purpose, and it was overseen by the Edison Company which was then quickly expanding the use of electricity around the country. To this end, the Spokane Falls Electric Light and Power Company was in business by 1885.[43] This was a steam powered lighting plant and it serviced some two-dozen street electric arc lamps.

What followed was a series of expansions, of electric service and power output. Power was offered to homes and businesses, and lines were strung down streets. The growth of the city, due to the booming mines in nearby Idaho, saw that there was hundredfold increase in demand for electricity. To facilitate this growth, more and larger generating stations were needed. At first a small hydroelectric dam was built on the Spokane Falls. Into this mix there were the streetcar companies who also needed electricity and built their own small generating stations.

The obvious solution to competing users and multiple generating stations was amalgamation, and the building of a large generating facility on Spokane Falls. (Spokane was blessed with two cataracts–an upper and a lower.) Strangely, the dominant Edison Company was uninterested in water power and it inflexibly stuck to its steam power plants much the same way as it later promoted direct current over that of alternating current. With this setback, and due to patent and contractual rights, a new company was formed. Thus the Washington Water and Power Company was set up in 1889.[44] The positive benefits of water power were suitably reinforced when a spark ignited a fire which demolished the city later that spring. Steam boilers and fires in the dry, pine country were and still are hazardous in summertime.

The rebuilding of Spokane only extended the customer base and electricity network. Thus the new company set about building a large 1100 kilowatt hydro plant on the lower falls at Monroe Street. "Work began on the powerhouse in 1889 and on the 18 foot high rock crib dam in early 1890."[45] The power plant was later expanded to 1400 kilowatts in 1894. It was a large two story brick and stone building.

With such output, the new company had corporate power: it proceeded to buy out the local Edison company and the street car companies. Byron Riblet was hired as chief engineer for the company and worked on the Ross Park extension. He went on to build forty miles of streetcar lines and supporting electric network in the early 1890's.

Despite some upsets with the 1893 economic depression, Washington Water and Power company went on from strength to strength. It later built larger dams on the Upper Falls, a large steam plant at Ross Park, and other dams at Long Lake and Post Falls.

While Byron Riblet left the electric company after a few years of service to conduct general contracting work, the private electric company is still in business today servicing the electricity needs of a large share of Eastern Washington and Idaho. It should be noted that this electric utility has a large, prominent brick building on the Spokane River which is a landmark in the city. Facing it is the Monroe Street road bridge, another famous feature in the Inland Empire City. And, in an interesting aside, travelling over all this is the "Spokane Falls Experience"chair-lift link; built by Riblet Tramway Company, with sponsorship by Washington Water and Power Company, for the 1974 Spokane Worlds Fair. It connects the northbank of town with the old fair site and continues to run as a tourist attraction. In the end it is fitting that the power company supported this flagship installation, and that the Riblet Company and the electric company still have connections, eighty years later, after the first electricity was turned on.

7. The Riblet Family

The name Riblet is of French Huguenot origin. The pre-history of the name are thought to have come from Robelet, Riboulet, and Riblette, meaning a rasher of pork. This is how the term 'riblet', for a pork product, was derived. During the religious disturbances in France during the seventeenth century, namely the St. Bartholomew's Day massacre in 1685 when Huguenots were attacked, religious minorities fled France. As a result, the Riblet family settled in the Palantine in Germany. In the process, a Christian Roblet later left the old world and came to America on the ship *Hope of London*. He arrived in Philadelphia in 1733. Eventually, he and his son Bartholomew settled in the Northern New England colonies.

While the records are few in number, what is know is that Bartholomew married in 1757 and is on record for petitioning to protect a village from marauding Indians. He also saw Revolutionary War service on the side of the Colonists. Later, he raised a son, also Christian Riblet, (b. 1761.) who eventually moved to Ohio.

One descendent of Christian Riblet was William Jackson Riblet, who was born in Ohio about the time of the Jackson Presidency.[46] William Jackson Riblet was a farmer all his life and he moved around in agricultural pursuits. He lived for a time in Wisconsin, where he married in 1861 before moving to Iowa. His offspring were born in Osage, Iowa. His wife was Annie Bell Riblet, (née Sutherland) and who was originally from Canada. Into their family there came three sons (in order by age): Walter, Byron and Royal. There also was a daughter named Nelly.

Byron Christian Riblet was born on February 20th, 1865 and raised on the family farm in Iowa as the second son. He became the cyclone around which the rest of the family orbited. The two brothers worked for him, and the family moved to the location of his field of work. Few details are known about the man as no diaries or letters have come to light. We are left with mainly secondhand accounts of his work. The few domestic and day-to-day facts that are known are as follows—Byron Riblet married Hallie Chapman on February 11, 1893. Hallie was the daughter of a distinguished Oregon family: her grandfather Colonel Chapman was a larger than life business man who helped organize the territory's first newspaper The Oregonian, he took part in the Cayuse Wars against the Indians, and later had mining interests in Montana. Hallie and Byron produced two daughters–Josephine and Virginia and eventually two grandchildren. Little information about Byron's family has surfaced, case in point no photograph of Hallie Riblet or her family have been discovered.

Byron Riblet later in life. Born in Iowa, his tramway work took him around the world from his Spokane base.

The elder brother Walter Riblet managed the Nelson office of Byron Riblet's Tramway Company and Walter lived there until his death in 1943. As the Kootenays provided much business for the firm, offices and engineering works were needed there. Tram parts were made in a machine shop and blacksmith shop, or contracted to the Nelson Iron Works. The Riblet Tramway Company first maintained offices in Sandon too around the turn of the century. Royal Riblet worked as a salesman for the company, and his role in the story will be elaborated on in its own chapter. After 1910 the firm gravitated to Spokane where Byron moved when mining activity in the Kootenay district dropped off due to fires and played out ore deposits. The

Nelson factory closed about 1912 and operations moved to Spokane.

About this time Byron Riblet bought a section of land on the Little Spokane River, near Colbert just outside the northern city limits of Spokane. He and his family lived there for the rest of their lives. In the process, he had the Spokane architect Kirtland Cutter design and build an attractive white, colonial-style house. It was large with almost a dozen rooms. Sadly, this house was destroyed by fire in 1933.[47] He then built a second home on the same property. Byron Riblet had an office on Main Avenue in downtown Spokane. His factory was originally on Division Street but subsequently moved to its present location on Grace in Spokane. In social circles Byron Riblet was a member of All-Saints cathedral, the Country Club and the elegant, plutocratic Spokane Club.

Unlike Royal, Byron did not leave behind a grand architectural or social legacy. Byron's fortune was largely consumed by the Depression, and he later sold his shares in the company to Carl Hansen. As a result, Byron has been relegated to obscurity. Byron died in 1952, while Hallie Riblet died in 1959. With Byron having no male heirs, the family name has, for the most part, died. (Royal Riblet did have a son, thus there is one branch carrying the name.)

As with the many filigrees to this story, now we have covered nineteenth century mining and engineering developments, Andrew Hallidie and the first tramways and mining in general in the 1890's, we can get on to the main branch—that of Byron Riblet and his mining tramways.

Notes —1

[1] K. Bittner. "Milestones in Ropeway History" International Seilbahn 1984. p. 184.

[2] See Thornton Wilder. *The Bridge of San Luis Rey*.

[3] T.E. Crowley. *Beam Engines*. Sire Publications Princes Risborough. p. 22.

[4] Sir Marc Brunel, another famous English engineer and who designed the plans for the first US Congress building, engineered machines to mass produce ship's purchase blocks in volume in 1799. It being the Napoleonic era, the British Royal Navy needed them in volume to equip their ships of the line. The experience of block rigging and manila rope in ships and their perfection of cable systems cannot be understated. Mariner's techniques were later applied to civil engineering, logging, mining and even rigging in the theatre. Sir Marc also invented the tunnelling shield on a project to bore under the River Thames at Woolwich. His ideas predated the methods used on the current Channel tunnel by almost two hundred years.

[5] See Schneigert. *Aerial Tramways and Funicular Railways*, p. 5.

[6] Stephen Johnson. *Encyclopaedia of Bridges and Tunnels*. Checkmark Books. NY. 2002. p. 250.

[7] William Randolph Hearst's father made his fortune in silver mines in Utah; apparently, W. R Hearst was nearly killed in an event, when a mine engine lowered a cage too fast in a shaft. Hearst was non-plussed by the incident.

[8] For details on the travellers and tramways of the Chilkoot (see AppendixVI.)

[9] German skullduggery and sabotage abounded in the First World War. The Zimmermann Telegram was only the icing on the cake: Germans blew up bridges (including the Cambie Street in Vancouver), attempted to blow rail way bridges in New Brunswick and the Fraser Canyon in BC, burned factories, destroyed barge loads of munitions in Puget Sound, and fomented trouble in Persia, India, and Ireland. The German spies also destroyed the Canadian Car Factory artillery shell plant in New Jersey. (See J.Witcover. *Sabotage at Black Tom*. pp. 80-81.)

[10] So extensive was the damage to goodwill and relations of the US and Germany by these act of sabotage that it stayed in memory long after the war while the reparation cases crept through the legal courts. Those cases

were settled finally by the Nazis. Even more oddly, the lawyer who prosecuted the case would later have his office in a World Trade tower. Ibid. p. 81.

[11] The abandoned mining towns in the West were impacted by these events as they soon saw life with some unwilling inhabitants. Franklin Roosevelt, after he signed the Executive Order to arrest and intern the Japanese-Americans, stated that he did not want to see any more Black Tom-type events. Thus the Japanese-American citizens were interned for a previous memory-regardless of how valid that idea was. Canada followed suit and interned its Japanese citizens in the Lardeau and Slocan towns of BC and other places. Ibid. p. 81.

[12] Captain Reginald Belknap, "Wire Roping the German Submarine." *Scientific American*. March 15, 1919.

[13] American Steel and Wire Company produced 23 million feet of wire for the barricade; A. Leschen and Sons-10 millon feet and Broderick and Bascom-4 million. In total 80 million feet were turned out by nine manufacturers. Ibid.

[14] Many Canadian first World War submariners were lost at sea, attacked by their own side, or encountered the minefields of both sides. The BC submarines CC1 and CC2 saw service in the Strait of Juan de Fuca and later the Atlantic Ocean.

[15] Another internee, from another conflict was Alvo von Alvensleben. He was a famous financier who brought German capital to BC at the turn of the twentieth century. The Hohenzollerns and other German Aristocrats placed their money, through him, into mines, mills and fish camps. It is outlandishly proposed that he built U-Boat pens at his fish cannery in the Queen Charlotte Islands to prey on the Alaskan-Yukon Coastal traffic and aide the Imperial German Pacific fleet-the very same fleet that BC Premier Richard McBride was so anxious about. Von Alvensleben turns up often in BC history backing mining ventures in the Kootenays, Stewart, land speculation in Surrey and the Wigwam Inn on Indian Arm. During the war he was arrested, his assets were seized and he finished his days in Seattle rather embittered about the British and their Empire.

[16] Richard Boyer. *Labor's Untold Story*. Cameron Associates. New York. 1955. p. 142.

[17] See Lukas. *Big Trouble* for a long explanation of the Idaho labour disputes.

[18] For more on William Haywood, (see Appendix VII.)

[19] For more details on Joe Hill and Ginger Goodwin (see Appendix VII.)

[20] Kahn. "Andrew Hallidie" California Historical Society. June 1940. p. 4.

[21] Ibid. p. 4.

[22] Ibid. p. 18.

[23] The First Nations performed many contract jobs in early BC, often providing essential roles, though they rarely get mentioned in the history books. (See T. Carlson Ed. *You Are Asked to Witness*. Sto:Lo Heritage Trust. Chilliwack, BC. 2001. pp. 115-116.)

[24] Ibid. p. 116.

[25] The large six feet diameter bull wheel had a large annular ring on the out side edge of the rim. Around the rim were placed vertical, two inch wide pairs of steel legs that extended slightly over the edge of the wheel and which acted as tongs. The bullwheel looked like an outsize gear or mill arbor with the legs placed radially around the wheel. As the cable tightened up-due to load, cable slippage, wind, or bouncing buckets-the cable pulled the legs slightly toward the centre of the bull wheel and in the process the leg-tongs gripped the cable more tightly. The legs acted in concert, though were independently mounted and had spaces between them. By the time of the expiration of his patent in 1895, other manufacturers were using similar designs. Bull wheels, like these can be seen scattered around the West in decaying operations and museums. (See Trennert. *Riding the High Wire Aerial Mine Tramways of the West*. p. 11.) (See Kahn. p. 19. for the description of Hallidie and his early devices.)

[26] Ibid. p. 1.

[27] Ibid. p. 17.

[28] Ibid. p. 19.

[29] Another cable car museum works as a curiosity in Lladanau, Wales. It is much smaller than the Market Street line of the San Francisco system. San Francisco's cable system demise was hurried by the Great Fire of 1906; the complete razing of the city by earthquake and gas-main fire provided a great opportunity to truncate the once extensive system. The great Post War modernization binge effected further closures almost to the point of complete extinction of the species. (See George Hilton. *The Cable Car in America*. La Jolla, Howell-North 1982.)

[30] The location of the mine high above the town caused other problems too. It turns out that the miners living in their bunkhouses had limited sanita-

tion; their untreated sewage emptied into the creek where it found its way into the Nelson camp's water supply. As a result there was a typhoid out break and many people died. Nelson moved its water intake in the end but it did not diminish the perils to pioneers.

[31] Garnet Basque, Ed. *West Kootenay: The Pioneer Years*. p. 44.

[33] Nelson pioneers recall the metal screechings and noise of the tram when it was in operation. Apparently, the tram could be heard all around the town and over the lake on still nights.

[34] Robert Trennert. *Riding the High Wire*. p. 26.

[35] There were many fraudulent scallywags about in the financial markets and the frontier. Early Rossland was filled with them and the Rossland mining camp circa 1891 acquired a reputation for fraud. Whitaker Wright was immortalized and anthropomorphized into the animal-formed rapscallian Toad in Kenneth Grahame's Wind in the Willows. Further character traits for Toad were borrowed from Oscar Wilde (of Reading Gaol-era) and the pompous Liberal MP Horatio Bottomly. (Consequently, the themes of recklessness and ruin come into the children story.) Kenneth Grahame served as Governor to the Bank of England and thus was close to financial dealings in the City. He also possessed a hidden desire for rural life which found form in his story and Grahame's retirement to Pangbourne on Thames. For information on Rossland (see Jeremy Mouat. Roaring Days. UBC Press. Vancouver. 1995. pp. 56-61.) And for tidbits on Toad (see Peter Green. *Beyond the Wild Wood*. Webb and Bower, Exeter, 1982.)

[36] Red Mountain is famous not just for its mine but also by its skiers. Skiing first came to BC around 1900 at Golden in the Rocky Mountains, due to their heavy snowfall. Skiing was used for transport and for sport. Soon skiing was taking place in Rossland too. Early races were held there and skiing entered the local culture. With the Hallidie mining tram, and soon the Riblet trams in action at mines around the Kootenays, it was not much of a diversion to create a ski lift. As a result, Cominco engineers cooperated and produced an early skilift at Red Mountain in 1948. It resembled the Riblet tramway, from where it took inspiration, with wooden towers.

[37] Lewty. *Across the Columbia Plain*. WSU Press. Pullman 1995. p. 27.

[38] The Northern Pacific Railroad signed a large leasing deal with the Pullman Railroad Car Company to share ownership of railcars. This is why that company was so honoured with a town named for them. (See Teri Hein. *Atomic Farmgirl*. Fulcrum Publishing. Colorado. 2000. p. 39.)

[39] Lewty. Op. Cit. p. 108.

40   The nineteenth century economy was still underwritten with money that was backed by gold. In the United States coins contained silver, thus the US Treasury was a large buyer of silver. (The mints of Mexico and China were other large consumers of argenta; India imported a sizable share for cultural reasons.) The unevenness of North American society in the latter half of the nineteenth century produced several social movements. In this manner the temperance, suffragette, progressive political parties, union, and proto-environmental movements were born. The radical West spearheaded most of these movements. Other outlandish notions included public ownership of utilities, controls on business, income tax on the rich, and "free and unlimited coinage of silver by the federal treasury." (See Carlos Schwantes. *The Pacific Northwest-an Interpretive History*. University of Nebraska Press. Nebraska, 1996. p. 263.) By 1890 these various forces coalesced into political movements.

The issue boiled over into the metal content of coins. The Sherman Silver Purchase act revived the practice of silver in coins. Idaho, a large silver producing state, was in favour of the Treasury producing unlimited quantities of silver dollars. Whereas, Eastern capitalists and manufacturers were wary of runaway inflation, and wanted to control the number of coins and to peg the greenback dollar to gold. Thus a political battle came to the forefront, and the issue was called 'bimetallism' after the fact that the dollar was valued against two metals. In the early 1890's the value of the dollar weakened, and the US gold reserves were reduced as a result. At the same time, profligate spending by the government further strained its gold resources. Part of this spending was the Sherman Silver Purchase Act that decreed the buying of tons of silver. With the election of the Democrat Grover Cleveland in 1892-whose party endorsed the Sherman Silver Purchase Act-the country was on a financial precipice. It only took the failure of one shaky railroad and everything collapsed. "Stock Prices plunged, the gold reserve plummeted, and by the end of the year [1893] 74 railroads, 600 banks, and 15,000 businesses had failed." Paul Boyer et al. *The Enduring Vision*. Heath and Co. Lexington, MA. 1995. p. 456. Into this mix, came the serious economic depression. Jobs disappeared and many were ruined. It was a dangerous political climate, so des perate were the people that some hijacked trains and headed to Washington D.C. to lobby the government with Coxey's 'Army.' Another issue was the collapse of farm prices, brought about by the overproduction with the new Western farms coming into production. In a word, the farmers were bankrupt.

One cultural legacy of this time of real misery is in the eerie iconography of Hallowe'en. The resulting bankruptcies of the farmers, banks and middle class caused people to either lose their houses, or physically abandon them. Thus myriads of Queen Anne Style and Second Empire design wood frame houses around America were left to the wind as people fled West seeking prosperity. The neglected houses eventually became empty and windowless-so by the 1920's they took on a sinister tone-and became connected with the ghosts and hauntings in the imaginations of young people. So prevalent were these empty homes that by the 1950's the image of the abandoned, Victorian house came full circle into a hallmark of the national psyche with the Addams family cartoons and houses at the Bates Motel.

These secondhand events aside, in the 1890's, things were in crises. In order to restore government finances, Cleveland repealed the Sherman Silver Purchase Act in 1893, but this one deed could not stop the flight of European capital from America. The downward cycle of the dollar continued. Only the expensive purchase of gold to shore up reserves reversed the trend. That gold was bought with discounted Treasury bonds by an avaricious J. P. Morgan and Wall Street. Ibid. p. 457. The political drifting and economic chaos ended Cleveland's term in office.

By clinging to the gold standard, Cleveland forced his opponents into an exaggerated obsession with silver. The genuine issues dividing rich and poor, creditor and debtor, farmer and city dweller disappeared in murky debate over two semimythic metals. Conservatives warned of nightmarish dangers in abandoning the gold standard and agrarian radicals extolled silver as a universal cureall. Ibid. p. 459.

And so the stage was set for the 1896 election.

William Jennings Bryan was elected to the Democratic nomination. He was a righteous and religious individual who had practiced law, and been elected to Congress. He was also a powerful orator. Splitting the liberal vote was the Populist party. This was a radical, Western third party that emerged in elections from time to time. The Republicans fielded William McKinley, who ran with Conservative, Establishmentian policies of tariffs, and the gold standard. McKinley, of course, won the election. He then went on to distract the nation from domestic issues with the Spanish American War.

The Klondike gold rush, and the re-emergence of wealth due to the sinusoidal nature of the business cycle, produced a rising market by 1898. Furthermore, the business consolidations or "re-Morganizations" with the control of railroads, mining and manufacturing into the the Rockefeller-Morgan-Mellon sphere furthered improved market conditions. McKinley passed the Currency Act of 1900 backing the gold standard. Under such conditions and popularity, McKinley again defeated Bryan in the election of 1900. However, the aggrieved working class was never dealt with. As a result tensions increased to the point where an alienated, unemployed anarcho-syndicalist, and Polish emigre named Leon Czolgosz assassinated McKinley in 1901. William Jennings Bryan was never elected president but went on to prosecute in the famous Scopes Monkey Trial.

41 A most sensible way of developing compared to the modern day whereby tracts of land razed, access put on via already overcrowded roads, strip development expanded, ugly boxes built and public transport forgotten.

42 Brill was a large maker of street railway cars. The 1947 issue of trolley cars on Vancouver streets were built by Brill. Many of these now languish in the swamps of Surrey. Quotation from C. Muetschler. *Spokane Street Railways-An Illustrated History*. Inland Empire Railway Historical Society. Spokane 1990. p. 16.

43 Steve Blewett. *A History of the Washington Water and Power Company 1889-1989*. Washington Water and Power Co. Spokane, 1989. p. 3.

44 Ibid. p. 4.

45 Ibid. p. 8.

46 "Historical Biographical Sketch of the Riblet or Riblett family" Cheney Cowles Archives.

47 John Fahey. The Brothers Riblet. *Spokane Magazine*. November 1980.

Cableways in the Kootenays

8. Sandon, B.C.

The mineral wealth of the Kootenays was beginning to be discovered by the last decade of the nineteenth century. As discussed earlier, the mines of Rossland and Nelson were returning large profits. To this end prospectors fanned out and began to search the region of Kootenay Lake. Mineral outcroppings were discovered at Ainsworth in 1889, which, together with the salubrious and scenic hotsprings, produced a frontier town. Ainsworth became a mining town in its own right, though it was largely overshadowed by its more famous and richer neighbour of Sandon.

The riches of the adjacent Slocan District were subsequently first unearthed by Andy Jardine, Jack Allen and John McDonald in August 1891 at Whitewater, a location some thirteen miles west of the future town of Kaslo, high in the Selkirk Mountains.[1] The reports of minerals produced a stampede of prospectors. One stampeder was John Seaton, who located the future Payne Mine, on Payne Mountain—a palisade type peak that defined the contour of Carpenter and Seaton Creeks. Another prospector was John Sandon, from whom the town would take its name.[2] Payne Mountain, Carpenter Creek, and Idaho Peak would form the backdrop for the fabulous wealth that the "Silvery Slocan" would soon produce.

In the ensuing rush the region became a veritable patchwork of mining claims. The names have since gone onto legend and history: Whitewater, Payne, Ruth, Wonderful, Reco, Sovereign, Rambler, Last Chance, American Boy, Noble five, Queen Bess, Ivanhoe, and Alamo are a few examples. With the mining output that these mines produced, the towns of Sandon, Three Forks, Silverton, New Denver and Kaslo sprang to life. With the mining came the impetus for building two railways and linking those connections with fleets of lake steamers.

Yet there were also some ups and downs in the story. The mines were first developed around 1892 with the hand clearing of trees, driving of shafts, and building roads and cabins. Materials were hauled in by sleigh and the ore rawhided out. (Rawhiding means to sack ore into a leather hide, and drag or pack out down the mountain.) Then the market collapsed. In May of 1893, the overturning of the Sherman Silver Purchase Act in the US precipitated a huge business depression.[3] With the US Government not buying volumes of silver, prices collapsed. As a result, nascent mines became dormant. To compound matters, the year 1894 brought widespread flooding around the Pacific Northwest. Towns were inundated and railways were washed out. It was the worst flooding in fifty years, and only comparable to the floods of 1948. If that was not enough, a fire demolished the young town of Kalso too.

Through it all, the high mineral content of the mines prevailed. Thus mining resumed in the Kootenay by 1895. Assisting matters were the fact the Great Northern Railroad laboured and built the narrow gauge Kaslo and Slocan Railway from Kaslo to Sandon and it was finished by the end of that year. The Canadian Pacific built its line in 1894 from Nakusp first into Three Forks, four miles below Sandon.[4] The CPR line to Sandon was finished in the next year. With easier shipping, industrialization was possible in the Slocan.

Industrialization meant machinery and some of the machinery needed was aerial tramways to move ore down from the steep slopes of the Slocan Mountains. First Hallidie and similar tramways were imported into the region but the new Kootenay mining business created an opportunity. Into this scene stepped Byron Riblet. Sandon became the mountainous maternity ward where Riblet tramways were conceived and born; later, the Kootenay Region expanded into a nursery where the Riblet trams shed their short pants and matured in. In the course of the following two-dozen years, Riblet produced about thirty aerial tramways for the immediate Slocan region. For tramways in the Slocan and Sandon region, Riblet was the supplier. With many of the mines installing tramways, and then rebuilding them as needed, the success of the Riblet tramway was guaranteed. Byron Riblet lived in Sandon from 1897 to about 1901 when he moved to Nelson. As we can see from the list of mines, he was called all around the Slocan and Kootenay region to install tramways.

By the last years of the nineteenth century, mines, mills and a communites sprang up in the mountains of the Slocan. The mines changed hands to investors and their pools of capital that allowed for huge frame mills, tramways and roads. Machinery was also imported and this leads us to a problem. Accurately determining the equipment installed at each Sandon mine is a near impossible task. With upwards of thirty mines and each spanning a period of more than fifty years, the levels of detail about mining are tremendous.[5] Complicating this issue was the temporary nature of the mining operators and the patchy records recording the events. After a discovery, the prospector usually sold the mine to a developer, who hand extracted some ore. Then this second person would again sell to an industrial concern that would then install some equipment and remove the known deposits of high-grade ore. This period usually lasted no longer than five years. After the easy riches had been removed, the mining stopped. Other lessees would come in, drill exploratory tunnels and possibly find more ore. If they did, the mining, milling and transport equipment would be upgraded. In this way several of the mines went through multiple concentrators and, in some cases, three separate aerial tramways.

The town of Sandon itself was also subject to this pattern of renewal and change. With the discoveries of 1892, a small mining camp with cabins scattered hither and yon was built. Compounding matters was the fact that there was little flat land in the valley. As a result, the town was built on the closest thing to flat land, the land beside Carpenter Creek. Despite the close presence of water, the town was nearly destroyed in a vast fire in 1900.

Sandon was rebuilt, this time Carpenter Creek was corralled and boarded over to form the main street. Larger and newer buildings were put up. Electricity was installed. Apparently, the prosperity of the town was secure and can be characterized by the records of the Hotel Licensing Act. At this time Sandon had 24 hotels, and 23 of them were licensed. The missing hotel license can be explained by the need to provide an evidently much needed 'quiet place.'

The Sandon Mines:

Noble Five

This famous mine was discovered in 1891 by "W.M. Hennessy, J.J.Hennessy, Frank Flint, J. McGuigan, and J. Seaton."[6] The group attempted to bushwhack back to Ainsworth over the mountain after a prospecting trip. In the process they discovered the mineral outcroppings. Each man staked a claim, and one more that was named for the group—hence the Noble Five. *The Report of the Minister of Mines* notes that the mines were worked "at a handsome profit."[7] Consequently, the Noble Five Consolidated Mining and Milling Company was formed in 1896 with J. McGuigan serving as manager. The mine itself is near the headwaters of Carpenter Creek, at Cody, three miles upstream from Sandon. Cody grew into a town in its own capacity. This included stores, houses and even a post office. The Kaslo and Slocan Railway ran five hundred trips into the town in 1896. In its time, Cody was a busy place.

With the railway into Sandon and a mile and a half spur to Cody, full-scale development was possible. The Company thus built a concentrator and steam power plant. As the Noble Five Claims were high on the steep, adjoining Payne Mountain, development was hindered.

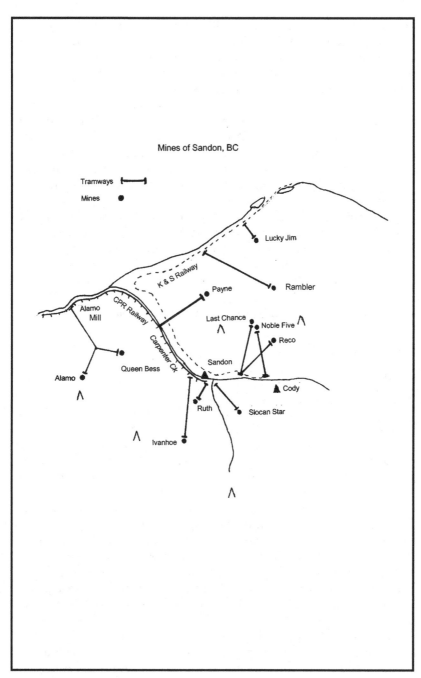

Sandon, B.C.

To improve transport a tramway was ordered.[8] When it came to install the tramway, the Noble Five Company advertised for an engineer in Spokane. Byron Riblet answered the notice, and thus reported for work in Cody where he became the Chief Engineer.[9] Despite the fact that he was inexperienced with aerial tramways, Byron Riblet was already a seasoned engineer and thus was ready to try his hand at mining tramways. One story that was circulated was that Byron Riblet misread the advertisement and mistook mining tramway for a street railway, which he had built in Spokane. In the end Byron Riblet installed the tramway as detailed to do. Yet he was not happy with its performance.

The tram had twenty towers supporting this heavy-industry high wire of $1^{1}/_{8}$ inch cable. It was 6100 feet in length. Two men operated the tramway: one to load and one to work the brakes.[10] In the process Riblet tinkered, and he experimented to make the tramway work better. The Noble Five tramway was the first Riblet Tram, and was in operation in 1897. Over the next five years Byron Riblet would experiment and perfect his line of tramways at several mines. One result of this was his numerous patents. (See Appendix II.)

The 1898 *Report of the Minister of Mines* cites that thirty-five men were employed at the Noble Five mine. Ownership of the mine was transferred about a half dozen years later, in 1905, to James Dunsmuir, the coal-mining magnate. Production dropped off significantly in that year. A dozen years later, with the rise of metal prices due to the First World War, Sandon experienced a period of resurgence. Many of the old claims that had been 'high graded' at first were then re-worked. About 1917 "a new cable tramway to Cody was under construction, as well as a new 250-ton concentrator" on the Noble Five mine.[11]

Mining continued through the years. In 1927 a haulage cable broke on a vertical man-lift cage and the elevator crashed to the bottom of its shaft, killing three men. The dead were transported out via the tramway for proper burial. Mining was a grim business. During the Great Depression mining activity in the valley was curtailed, though the re-establishment of the gold standard helped bring some life into the community. "The mill, powerplant, office and lower tram terminal were completely destroyed by fire" in 1944. Yet this was not the end of the mine.

A new company the Cody-Reco Mines Ltd. was formed and assembled the old Noble Five, Sovereign, and Reco claims. It built a new concentrator at the Noble Five site in 1952. The ruins of this building stand to this day. Below the concentrator is a derelict Fairbanks Morse, Model "Y", two-cycle, semi-diesel engine that pow-

Noble Five Mill near Sandon, B.C. at Cody about 1898-it was the site of Byron Riblet's first tramway. NAC PA 17833

ered an electric generator.[12] The Noble Five operations were rebuilt with a new concentrator and powerplant and were back in operation by the following decade. Numerous lessees worked the claims over the next decades for an eventual, total yield of 15 tons of silver, 1588 tons of zinc, and 2173 tons of lead.

Last Chance

Two thirds of the way up Carpenter Creek, on the southwest flank of Payne Mountain was the Last Chance Mine. It was high up (6300 feet) on the avalanche slope above Cody. The claim was located in 1891.[13] Development work continued through the 1890's. The Last Chance mine was adjacent to the American Boy claim. (A mine that was worked by the American Boy Mining and Milling Company from 1897 and lead by Tom McGuigan, the pioneer of the district.)

An aerial tram was built to access the Last Chance and American Boy mines and was "about completed" in 1898. There was an ore bin on the Kaslo and Slocan Railway spur to Cody, just before it crossed the large bridge over Carpenter Creek. The mine had forty-five men on the payroll and shipped 1700 tons of ore in Queen Victoria's Diamond Jubilee Year of 1897.

Payne Mine

John Seaton and Eli Carpenter staked the Payne mine in 1891. It was really a cluster of claims with the Maid of Erin, Two Jacks and Mountain Chief as part of its mountain halls. One A. McCune bought the mine in 1896 and worked it profitably. He subsequently sold it to Payne Consolidated Mining Company in 1900.

There were extensive mining operations at the Payne. Eleven different levels of tunnels were bored, and some were drilled all the way through the mountain. To handle the ore an impressive inclined rail tramway ran 5500 feet up Payne Mountain. It had a capacity 500 tons per day. In 1902 a concentrator was built at the lower terminal. Also, interestingly, "a wire rope tramway ran to the CPR tracks below."[14]

The Payne mine had one hundred-thirty men working in its operations. In 1898 the Payne shipped 14000 tons of ore. Most of the work was undertaken between 1898 and 1904 with the high value ore stoped and removed. Large snow slides fell in 1904, near the Payne mine causing trouble and thus mining ceased. Lessees worked it thereafter. The Payne Mines were formed in 1907 and they erected a 120-ton capacity concentrator. The 1910 conflagration, fueled by the wantonly destroyed, dead and desiccated forest, burned not only the Kaslo and Slocan Railway but the Payne Mine equipment and buildings too. The large forest fire stopped the first wave of development in Sandon. Only the rise of metal prices at the end of the First World War provided the impetus for renewal in the region.

A new company, the Slocan Payne Mines took over the operation of the Payne; a new aerial tram was built to replace the burned funicular and provide access the mines on Payne Mountain. A map in the 1922 *Report of the Minister of Mines* notes its location. Sporadic work continued at the Payne in the inter-war years; the mine produced 50,000 tons of ore grossing some 5 million dollars over its lifetime.

Reco

A man named Ruecau located this claim near the Noble Five in 1892. He teamed up with the Sandon fixture of Johnny Harris and registered the clearer spelled claim the Reco. Development continued and by 1898 "a tramway and millsite had been surveyed." The tram was eventually built by 1900 and it ran diagonally to the slope at an east-west orientation. This tram crossed the Last Chance Tram. Alas, there is very little information on this tramway other than it operated for about eight years after the year 1900. Other operators came in, and presumably used the linkage. The Reco was later amalgamated into the Sovereign mine.

Sovereign

This mine was staked early on and first shipped ore in 1899. It, too, was near Cody. By the turn of the century, the Sovereign had over 2000 feet of tunnels. After such interest, the next fifteen years was a period of desultory digging done by lessees. That was the case when Clarence Cunningham optioned the property in 1916. With a stroke of luck, and some twenty-five feet of exploratory tunnels, Cunningham discovered a rich silver lode at the Sovereign. To harvest this windfall, Cunningham erected a 7000 foot tramway.

Ruth

The Ruth claims were located in 1892. The claim was hand worked for the interim period. Development was initiated and the operators contemplated to put in a mill and tramway in 1898. Joining forces with the nearby Ivanhoe claim, a tramway and mill was constructed in 1900 at the confluence of Carpenter and Tributary Creeks. Byron Riblet built the tram. This mill burned down in 1914. Obviously the tramway was rebuilt thereafter, and the remains of this tramway can still be seen on the steep slope to the west of town.

Richmond-Eureka

Due south of Sandon, near the hydroelectric plant, was the Richmond-Eureka claim. Davis, the mining engineer, recollected:

> In the spring of 1907, I went into the Slocan country to open up Cominco's Richmond-Eureka mine, lying right about the town of Sandon. It was a little more than a prospect at the beginning, but we soon opened up a nice little ore shoot of high grade silver-lead ore which netted us around $10,000 a month for nearly two years. We never had any power installed there, so it was hand steel all the way through [hand drilling]. An aerial tramway, four thousand feet long, installed shortly after my arrival, landed the ore right on the railway at Sandon and a sweet little operation it was. Nearly forty years on, on revisiting Sandon. I noticed that the right of way on this tramway was so grown up with new timber that it cound not be seen at all, and all that could be seen of the lower terminal was a pile of old and rotting timbers.[15]

Davis later became a respected engineer for Cominco and worked in the Stewart Region where he had a serious encounter with a grizzly bear.

Ivanhoe

The Ivanhoe was an early mine that was obscured by later developments. The Ivanhoe was a big operation and was first owned by the Minnesota Silver Company since 1896—with a Phillip Hickey serving as local manager. Large amounts of development work was done to 1898, yet full scale production was hindered as the "difficulty of approach has been an obstacle."[16] As a result no ore was shipped in that year. Management contemplated the building of a mill. In due course a mill and tramway was constructed and output increased. In the ten-year period of ownership $190,000 worth of ore was shipped by the Minnesota Silver Company. By 1904 some 3500 feet of tunnel had been driven at the mine on eight separate levels.

> The general ore chute is, approximately, 360 feet long in the vein, and extends down the No. 8 level.... On No. 3 the ore is found on both sides of the zone, and...made irresistible the temptation to extract it, and this was done, and the result that the filling of the fissure was undermined and began to show a tendency to slip down. This has been headed off by the timbering of a most substantial character, placed with great skill and no little risk.[17]

The ore was argentiferous galena.

To haul the ore "a Riblet aerial tramway, 8500 feet long" transported the ore to the Ivanhoe Mill. The mill was constructed soon after 1898 and used Blake Crushers and a separating belt, where the clean galena was picked by hand. The mill treated 70 tons of ore per day. And the ore contained about 55% lead and about 90 ounces of silver per ton.[18] The concentrates were loaded directly into Canadian Pacific Railway boxcars. Conveniently, the Sandon railway line ran immediately adjacent to the Ivanhoe Mill.

Queen Bess

Started in the year 1892, the Queen Bess had important ore reserves. It shipped over a million dollars worth of ore in its first days. Located to the Northwest of Sandon, it was above the Ivanhoe mine but east of the Alamo claims. A feeder tramline connected the Queen Bess mine to the Alamo tramway. They were part of Cunningham's large and profitable operation. The ores were then treated at the Alamo Concentrator. The mine was worked for twenty years until the First World War when the companies moved on to seemingly more productive bodies. Thereafter, the Queen Bess operated by small lessees until after the Second World War.

In 1948, Viola MacMillan bought the mine and contracted the previous owner to explore for new ore. Soon rich new silver veins were discovered and the mine sprang back to large scale production. In time, Violet MacMillan became the Grand Dame of Canadian Mining and was successful in a largely male occupation.[19] The Viola Mac mine worked for ten very productive years, delivering 24,000 tons a year. A large waste scree pile marks the mine location on the hillside and can be seen high on the right from the road into town.

Alamo

The Alamo mines were situated high on Idaho Peak and this presented access difficulties. As the mines were in the alpine, on the high ridge west of Sandon, getting there via Sandon was difficult. The Alamo and Idaho Claims were staked by H. St. John, and E. Gove in 1892 during the initial rush. Hand development continued and the concentration of the ore determined that a concentrator would be built

> Near the junction of Howsen Creek and Carpenter Creek. The concentrator was completed and throughout 1895 they were treating twenty tons a day averaging 153 oz silver to the ton.... The tramway from the mine to mill was in two sections 3400 feet and 3375 feet.[20]

As this pre-dates Byron Riblet's arrival, this was probably a Hallidie tramway.

With the equipment the owners—the Slocan Mining Company—were pulling 50 tons of ore a day. The mill was then sold to a Scottish Company the next year, and they removed a further 4000 tons of ore. Seven years later, in 1903, new owners took over the property and improved the mill and tramway.

> A Riblet aerial tramway, 6000 feet long, with a rise of 3000 feet, runs from the mill on the railway up Howson Creek to the junction of the Alamo and Idaho basins, and can be produced in a straight line to the Alamo workings, if desired. In the meantime, a "baby tram," 1800 feet long, runs from the upper terminal of the main tram to the Idaho workings, and over this all the Idaho ore is hauled.[21]

In the process the mine shipped over a total of a million dollars worth of ore. The Idaho claim ore content was 105 oz of silver per ton, while also containing 55% lead as well. It was found that the soil atop the region was laced with lead and silver, to a content higher than that in the vein, and thus was subsequently removed.

The Riblet tramway delivered the ore to a lower terminal with a 1000 ton capacity ore bin. In turn, the Alamo mill had a 100 ton daily capacity. To process the ore "the plant consists of crusher, 3 sets of rolls, Huntingdon mill, 2 elevators, 10 jigs, spitzkasten, 4 Wilfleys, and 2 extra Wilfleys to treat tailings from the first four."[22]

The Last Chance mine, just south of Sandon. It was first staked in 1891 and the tramway was built after 1898. The Kaslo and Slocan Railway ran below the mine onto Cody Photo: BCARS C 06758

As with so many mines, for the next dozen years the Alamo lay fallow. In turn, Clarence Cunningham, flush with success from his Queen Bess bonanza, reinvested his money into the Idaho-Alamo operation. Cunningham bought the mine in 1916. He reinvigorated mining in the region with the building of a new, quarter million dollar concentrator at Alamo on Carpenter Creek. Presumably, the tramway too received some attention from Byron Riblet—our wire rope wizard though it is not stated in the records.

Rambler

This mine was sited in the McGuigan watershed, on the backside of Payne Mountain from Sandon. Staked during the early rush, it was in operation by the turn of the century when it profitably produced for 34 years. Many levels of raises were thrust into the mountain in the successful location of silver ore. Obviously, over such a span of time and scale of operation machinery and a mill were needed. The first workings were demolished in the great forest fire of 1910. In time the plant was rebuilt. Again, there is no specific mention of a Riblet tramway at the Rambler-Cariboo mine though a tramway did work there. There is mention of the tailings being "trammed" out to be reworked at the Zincton mill during the search for gold in the depression.

Zincton-Lucky Jim

The Lucky Jim mine was a latter addition to the region. Located across Seaton Creek from Zincton, the Lucky Jim Mine sat on the south hillside. Its remains can be seen from the highway today on the road to Kaslo. Staked by a fortunate James Shields in 1892, it shipped silver in the following dozen years but later became a zinc mine due to the large zinc content in the ore. The tremendous forest fires of 1910 destroyed the Kaslo and Slocan Railway, the Lucky Jim mine, the Rambler Mine and the Payne Mine.

In 1910 George Hughes made the epic run for his life eastward down the Kaslo and Slocan Railway grade when the forest fire came sweeping up the valley from Three Forks toward the Lucky Jim, feeding on dead timber standing from a previous fire. Some men who dodged for safety into a mine shaft to escape the searing heat weren't so lucky as Hughes; the fire was so intense that it consumed all the oxygen and the men suffocated to death in the shaft.[23] While "the losses in this fire also included a new aerial tramway from the portal of No. 5 tunnel, down approximately 600 feet to the ore bins at the railway."[24] Later, the mine plant was rebuilt, and the CPR extended its railway to replace the lost narrow gauge line.

New companies bought and worked the properties, and the owners had successful years into the 1920's. With this, new wooden buildings and plant were erected. Only the collapse of the markets in the Depression idled the operation. The Sheep Creek Mines bought the Lucky Jim in 1937 and they brought new life to the operation. New ore shoots were discovered, and the payroll expanded to over fifty. It appears the aerial tram was rebuilt—Royal Riblet drew a 1947 blueprint for the Zincton tram—suggesting that Riblet built the original installation.[25]

One anecdote from the late 1940's was a worker named Andy Jardine who "was riding the timber bucket down the tram from No. 5 level. It broke down (the power went off) and he was between towers 60-80 feet above the ground, and it was minus twenty (F) at the time. Being wet and tired, he was very fortunate in attracting the attention of some fellows down below."[26] In turn the ground crew rescued him from freezing. The Sheep Creek Mines at Zincton stopped operation in 1953 though other people have worked the property off and on.

Sandon Town

The town of Sandon was the mining mecca in the Kootenays. As told each mine went through various periods of work and employment. Interestingly, during winters in the 1920's when the mines, tramways and town was flourishing, Scandinavian miners rode the aerial trams to the mines with their skis, then skied down from their mine to the town after work. In effect, this made the Riblet aerial tram the first ski lift, though no one thought much of it as it was part of life in the Kootenays. Business in the Slocan quietly slowed down during the Depression and the region began a period of decline.

During the Second World War the town's buildings were painted and fixed up by the Japanese Internees who were forcibly moved to the largely empty town. They worked on the buildings as something to do. Alas, paint and elbowgrease could not change economic cycles and the closing of the mines finished the town. Sandon was only abandoned after 1952 when the post office pulled out. Unfortunately, in 1955 a torrent of water brought debris down Carpenter Creek that blocked the culvert under Sandon's streets. The rushing water then flowed into town and demolished many of the empty, remaining buildings.

Some of the Bakke family en route to a mine at Sandon in a Riblet bucket.

Subsequently, the civic portion of the town was forgotten. Then came a period of decay: with no one around weather, thieves and vandals despatched many of the surviving town features. Sadly, the heavy snowload, damp, decay and fire finished much of the remainder. In spite of this, historic preservation, trails and public interest resumed in the 1980's. With increasing mobility, paved roads, and exploding population in the West, people began to revisit and inhabit out of the way areas. The Kendrick Block in Sandon was refurbished in 1997. A museum established, and cabins built. In short, life returned to Sandon. The present owners are very ambitious with large building and machinery restoration projects, historic research, and commerce established. It is good to see. Perhaps the installation of some modern Riblet skilifts would further invigorate the region into another mining camp and winter resort. If one descends on skis west from Idaho Peak above Sandon, one enters the Four Mile Creek watershed and the nearby mining town of Silverton. That is where our trail now takes us.

9. Silverton, B.C.

The other major mining complex in the Slocan District was the Silverton area. In essence, the Silverton mines were connected to the Sandon mines as they abutted the same orebody under the western flank of Idaho Peak and adjacent ridges. The town of Silverton lies on beautiful Slocan Lake, downstream roughly a dozen miles from Sandon on Carpenter Creek. As with Sandon, there were numerous mines in the immediate Silverton area—mines that all contributed in their way. These mines included the Van Roi, Hewitt, Wakefield and the Standard.

Slocan Lake allowed for access and easy communication to Nelson—south via the Slocan River—and north to Arrowhead on the Arrow Lakes. Indeed, the Canadian Pacific Railway eventually accessed the region from the northwest via lake steamers and a railway to Sandon. While most of the traffic journeyed to Sandon on the Carpenter Creek route, the mines of Silverton were located just to the south on Four Mile Creek—a local tributary.

Wakefield Mine

First staked in 1897, on Idaho Peak at approximately 5000 foot level, this mine was accessed via a three mile long road on Four Mile Creek. After that a trail wound up the hillside to the mine at elevation. A one-hundred ton mill was duly installed at Four Mile Creek in 1900 to process the ore.[27] To move the ore an aerial tram connected the mine to the mill. Two years later, development ceased and the operation was idled. Lessees worked the mill thereafter. In 1912 the mill burned completely, but was rebuilt as a complex of mines, including the Hewitt and the Wakefield, needed a mill in the area. There is no detail on who built the tram originally or assisted in its reconstruction after 1912.[28]

Hewitt

This mine was started during the early rush in 1893 together with the Lorna Doone working. It was hand worked and developed for ten years until a New York operation leased the mines. In 1907, they erected a 5000 foot aerial tram from their No. 3 tunnel to connect to the Wakefield mill at Silverton Creek.[29] Mining supplies, timber and food were shipped in via the tramway. Other interests then took on the property. They connected the Hewitt workings with the Lorna Doone Mine and camp via adits in the hill. Another aerial tram was

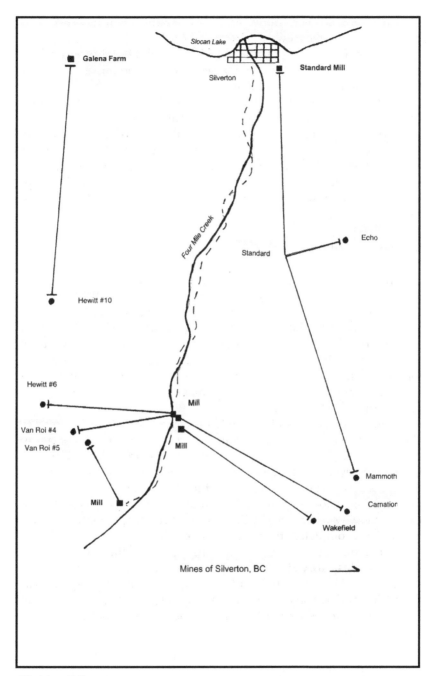

Silverton, B.C.

built from the adjacent outlet of the No. 7 adit down the Wakefield mill too. This larger Hewitt enterprise operated until a fire at the mill in 1912. After this a larger 150 ton mill was built, but it was not fully running until 1915. By this time lessees worked the mines, including Clarence Cunningham of the Alamo mine.

In the twenties other people optioned the site; investment later further increased when the Galena Farm mines took over the property. They drove a mile long adit through the hill to the western side where they built an ore bin. From there, a Riblet tram stretched one and one half miles down to the beach at Galena Farm where there was a mill. Numerous other operators have since optioned the property, done exploratory work, and removed some ore in the intervening half century.

Carnation Mine

In the early rush to the Slocan, prospectors staked much of the ground in the hills. Two mines near Silverton around 1892 were the Jenny Lind (after the singer), and the Carnation. Donald Mann, the investor and industrialist, owned the Carnation Mine and it shipped several pounds of ore in that decade.[31] By 1904, they were leased by the owners of the adjacent Wakefield Mine into their mining group. On the whole, little work was done on the property for 15 years.

After 1917, "considerable development work was done in No. 2 and 3 adits."[32] Then, in 1925, the English Victoria Syndicate bought rights to the property and proceeded to drive a half-mile of tunnel to the western side of the mountain. From there, a 9000-foot aerial tramway linked the No. 2 adit with the Wakefield mill on Four Mile Creek. However, due to the expensive and exploratory nature of mining, no large ore shoots were found at the Carnation Mine and mining stopped. The random sampling method of driving tunnels was an expensive way to prospect. Then the market collapse in the depression of 1930 completely finished the project.

Other players with money to burn optioned the property after 1939, and desultory digging was done off and on for the next 25 years. Some 300 kg of silver and 20 tons of lead and zinc apiece was found in the 500 tons of ore was eventually removed, though not enough to make the Carnation Mine a bouquet for the balance sheet.

Standard Silver

A patchwork of claims were staked in the early years just prior to 1900 above Silverton, these included the Emily Edith and the Alpha. These workings provided for some rich, localized veins of silver. After the initial material was removed, more tunnels were drilled and galena found. A 7900 foot Riblet aerial tram was built to connect the lake front mill to the Standard Mine.[32] It was rated at 20 tons per hour capacity. That mill was built in 1911. A 3000 foot jig back tram also connected the Echo Mine with the Standard mine. The mines were amalgamated into the Standard-Silver Lead Mining Company in 1912.

The Standard tram and mill became the nucleus around which the town of Silverton formed. In addition, to the mill, a hydro station, boiler house, machine shop and other work related facilities—house s , h o t e l s   a n d orchards sprang up. The Standard-Silver Lead Mine was an important mine for the next ten years. Unfortunately, the explosive political climate in 1919 created a situation that produced a strike at the mine. Reading the political climate, the One Big Union called a strike.[33] As a result, the mine never recovered from the disruption of the strike and the mine closed in 1921.

The easy way up the hill: two boys on a Riblet hook bound for the Standard Mine at Silverton, BC c.1918; normally riding was forbidden.

Mammoth Mine

First discovered in 1921, the Mammoth Mine was worked by several operators and lay fallow for a few years until 1929 when promoters came in and reconfigured the machinery. The mine was situated high on Idaho Peak above the Standard Mine. In 1929 the Standard tramway was extended to a final length of 14000 feet up Idaho Peak to the Mammoth Mine. In the process, a new mill at Silverton was also installed. Alas, with impeccable timing, the mine came online in 1930 and collapsing markets. As a result, the mine, tramway and mill operated briefly as a test, only to shut down while waiting for better prices. The mine worked again in the period of 1935 -37.

Interestingly, the mine contained retrievable ore and mining continued through to 1942. Eight years later, more work ensued and the tramway was back in operation until 1959.[34] Thus, the mines of Silverton, while separate from the mines at Sandon, are a continuation of the Kootenay Region's mining legacy. Byron Riblet continued establishing his name by engineering these successful mining tramways; he was asked to build them to other Kootenay mines south of Nelson and this is where the next chapter leads.

10. Moyie, B.C.

As underground mining activity increased in British Columbia—first with the Silver King mine at Nelson, then the mines at Rossland, Sandon and the Lardeau in the West Kootenays—prospectors investigated other areas of the province. Eyes were cast at the neighbouring East Kootenay region. The East Kootenay is a natural convergence point as the Rocky Mountain trench forms a geologic north-south route. In the Rocky Mountain trench is Columbia Lake—the headwaters to its namesake river—as it drains into the Columbia River winding its way through BC and fame in the US Pacific Northwest. Native peoples used this valley for millenia then later came the furtraders, catholic missionaries, and entrepreneurs. At one point a canal was attempted to link the two watersheds of the Columbia and the St. Mary's River at what is now Canal Flats.

One Father P. de Coccola was ministering to the local Natives at the St. Eugene mission on the same religious-toned river. There he ran the Mission, hospital and school. However, as the diocese was depopulated, operating funds for that eternal enterprise were slight in the late 1880's. To bring in revenue, Father de Coccola encouraged his flock to look for 'money rocks' or gold bearing quartz. Eventually

"Pielle" Tete de Fer returned from the field with a sackful of rocks. After having the rock samples assayed in Spokane, where the reports told of silver and lead, de Coccolla decided to stake the claims.[35]

The rocks originated from the MacGillveray mountains that separated Cranbrook from Creston. At the pass of those peaks lies Moyie Lake. It is a small, six mile long, crescent shaped lake defined by the mountain which rises steeply from its south shore. Atop that mountain was the motherlode. Three claims were staked—the Society Girl, St. Eugene, and Peter—and registered in 1893. Tete de Fer and de Coccola sold their interests to John Finch, an owner of Rossland's War Eagle mine, and took their proceeds to build a house, hospital, and church at Saint Mary's.[36]

At Moyie, the mines were consolidated into the St. Eugene group and development work started. By 1897 miners were hacking at the eight foot wide vein of galena. First the ore was rawhided and hand shifted to the wagon road at the Moyie Lake. There, wagons hauled it to the Rocky Mountain Trench where river steamers would haul it to smelters in Montana. With the rising mineral traffic, the Canadian Pacific Railway became interested in the region and pushed a railway line through the Crowsnest Pass onto the burgeoning Kootenay. The famed Foley Brothers were the railway contractors on the rail project.[37] That railway passed on the south shore of Moyie Lake as it does to this day. In 1899 Finch sold two thirds of his interests in the Moyie Mine to the Gooderham-Blackstock interests of Toronto who proceeded to industrialize the operation. Rock drills were brought in and a 300-ton per day concentrator was built at the lake. Water flumes were built and a town and bunkhouses constructed.

As the mine was atop the mountain and the railway at the lakeshore—an aerial tram was needed to connect the two. Byron Riblet was contracted in 1899 to build a "3300 foot long tram from the upper adit of the mine to the concentrator."[38] The tram headed from atop the ore bin at the concentrator, up a cut in the trees to the top of the mountain. The *Report of the Minister of Mines* elaborates that:

> while the buckets run as in the Otto or Bleichert Systems, the different being in this case the buckets are attached to the hauling ropes by an arm bolted to the rope, instead of being held by a grip. Each bucket holds 900 pounds of ore and is self dumping, being emptied as it passes over the bin by an automatic device without the necessity of being stopped, while the loading at the top bin is also done by an automatic loader.[39]

One man operated the tram.

At this early stage of development Byron Riblet, as stated, used a modified Hallidie bolt grip system. The bucket was connected to links woven into the traction rope, to which the buckets were bolted. He later abandoned this system and went with a spring-operated grip when he needed the ability to grip and release such as at an angle station. Riblet used Hallidie-style pyramid towers to hold the track and traction cables, and these towers were constructed of site cut trees.

The mine was rich and produced 115,000 tons of ore and earned $1.7 million by 1901. Yet in July of 1901, when confronted with falling metal prices, the mine reduced production. It was closed in September of 1902. Title was later regained by the Gooderham-Blackstock syndicate and they resumed production yielding "28 tons of silver and 18,250 of lead" during 1905. The St. Eugene mine was then sold to the Canadian Pacific Railway which formed the Consolidated Mining and Smelting Company (later Cominco) with the smelter at Trail and several mines in that organization's set of assets.

The Consolidated Mining and Smelting Company ran the mine for five years and was getting good production in that time. Its payroll had some 450 people on it. After 1908 with the economic "Panic of 1907" and declining ore, they decided to reduce the payroll. This was done gradually until the ore played out in 1913 when the St. Eugene mine closed. Other mining lessees continued to look for ore and workrd existing adits. Moyie's troubles increased in 1921 when the concentrator burned. There is no record of the eventual fate of the Riblet tramway. It was most probably dismantled and moved to another mine as trams were needed at other operating mines in the Kootenays. Today there is very little at Moyie other than cement foundations and a tailings pile. The tram slash can be seen on the hill but the towers disappeared long ago.

11. Ymir, B.C.

Ymir is on the Salmon Valley eighteen miles south of Nelson. That valley forms a natural corridor for travel. The Hall Brothers travelled up the valley, and prospected around the area, when they scouted what was to become the Silver King Mine (q.v.). In the process of the Kootenay mining boom, the Great Northern extended its rail lines—in the form of the Nelson and Fort Sheppard Railroad from Northport—Washington to Nelson, with a station and siding at Ymir.

Ymir's colourful name stems from Norse mythology. Ymir was a proto-god who was "slain by Odin and his brothers Vili and Ve and out of his body created the World. Ymir's flesh became the land, his bones the mountains, his blood the lakes and streams."[40] The Viking allegory is apt considering the mountain and wood setting and the town's current outlaw undertow.

At Ymir's outset, there were many mining prospects in the region—Dundee, Hunter V, Highland, Fern, Arlington, Spotted Horse and Tamarac for example. Later, mining consolidated around three main producers at Ymir: the Yankee Girl, Hunter V and the Ymir mines.

Ymir Mine

In 1900, the Ymir mine had a 2400-foot Hallidie tramway which could carry 250 tons in ten hours.[41] Though the Report goes on to say that the tram was "not kept up." Other difficulties with this tram were referred to as well.

> [The] introduction of a better loading device and provision of larger buckets. Operations show a good savings, the cost being $7^{3/4}$ cents per ton, while in 1898 it was $15^{1/2}$ cents per ton. The difference, however, was largely due to the greatly increased tonnage...[and] includes the cost of 2 new cables, one of which proved to be of very poor material and had to be discarded after carrying 2000 tons.[42]

It would seem that the Hallidie design was operating at its limits again. Whereas, the lower tram operator once did not operate to his limit. The Hallidie tram had smallbuckets on hangers and they needed to be tripped, to allow the hopper to empty into the bin. Instead of using an arm and shoulder, the operator realized he could unlatch the bucket with a swift kick. This worked well until one day his pant cuff caught in the catch mechanism—he then was catapulted away and inverted like a pair of trousers pinned upside down to a clothesline. Only the quality of his work pants and the quick action of another operator saved him from headlong doom.

The Ymir mine "was by far the most important mine in the region with 80 stamps."[43] Each stamp could crush 2 to 3 tons of ore per hour. Steam engines powered the stamps, though hungry boilers were needed to generate the steam for those engines. The boiler furnaces consumed one cord per hour of wood. As a result, the hillsides were stripped of trees to meet the demand.

Ymir had other aerial trams in operation too. The Tamarac Mine had "a wire rope tramway" one and one half mines in length. It had a 75 ton per day capacity, but it too had "tramway troubles" of an unspecified nature. The Highland mine also had a Hallidie tram.[44]

Hunter V

A big ore producer in Ymir was the Hunter V mine. Situated on the south side of Porcupine Creek, the ore body was large and mined by 'glory hole' method or removing a large cavity.

> **The ore was then loaded into the buckets of a short aerial tramway for delivery to the bin at the head of the main tram. From there it was delivered to the ore bin on the Great Northern Railway at Porcupine Creek siding $2^1/_2$ miles away and about 200 feet lower in elevation.**[45]

The main tramway was built by the Riblet Tramway Company; and it was in operation by 1904. There were numerous mishaps with the tram.

> **On one occasion, the mine superindent was riding a bucket on the upper short tram. The bucket was dumped in to the ore bin owing to the operator failing to stop [the bucket and boss] before the dump. Luckily, the bin was full of ore and the man's fall was only a few feet so he wasn't hurt. When told that he had dumped his boss into the bin, the operator said "That will learn him to be smart."**[46]

Other incidents were ice on the bull wheel and cable grips, which allowed the buckets and cable to run dangerously free. Another time the traction rope came off the control sheave at the upper station causing a long delay. As it was refitted into place the tram stopped. Unfortunately, there was a passenger hauling lumber on the line at the time, and rather than be stuck, he decided to dismount using a rope. The slide down the rope caused serious rope burns to his hands.

Additionally, a man named Norcross operated the tram and he is welcome source of information about the tram. His son reminisced on a few details about his father's workplace.

> As small boys we used to go to the tramhouse to watch the operation. On one occasion my father looked behind him to see my brother Albert ten feet in the air with his mackinaw coat accidentally hooked on the counterweight of the automatic loader. As the passing bucket released the loader lowered him to the floor. [47]

The Hunter V mine closed down in 1906.

Yankee Girl

The Yankee Girl mine was another large producer in Ymir, and one that had longevity. Staked in 1899, the claims were 2500 feet

Looking down the Riblet Tram to the Yankee Girl Mill-note how it continues over the Salmon River to the railway. Ymir town is in the background in this photograph from the 1920's. Ymir is just south of Nelson in the Salmon Valley.

above the town. Various operators extracted ore over the next decade. New lessees brought in modern equipment in 1911. This included a "6000 foot aerial tram" and enlarged mine plant. The tram had seven towers and one large span of 2000 feet. It brought the ore from the mine mouth above town, down to a terminal and bin on the East Side of the Salmon River. The ore was then concentrated and reloaded onto a short 500 foot extension of the tram, over the Salmon River, to another terminal, bin and railspur on the Great Northern. And "during the period named (1911-1919) 16,500 tons of ore were shipped averaging 0.82 ounces of gold per ton."[48]

Various operators mined the Yankee Girl on and off. A new mill was built in 1934 for resumed production. That mill survived for many years, though its visible situation in adjoining the town in full view made it the target of vandals. The machinery was eventually scrapped, though the cement foundations can still be seen.

12. Porto Rico, B.C.

Almost halfway between Ymir and Nelson is the long Barrett Valley. Seven miles up that valley—from the main Salmon Valley—is the Porto Rico Mine. It was discovered by two men—Maxwell and Day—in 1896. The discovers sold the prospect to an English concern who began to drill for ore. Seven hundred feet of tunnel was driven in 1897. "In the same year, 41 tons of ore, sent to the Trail smelter was reported to have yielded $76.25 per ton in gold."[49]

Consequently, a 10 stamp mill was installed as was a 2500 foot aerial tram. "The aerial tramway was installed by Mr. B.C. Riblet, then of Sandon, Slocan District."[50] It was in operation by the end of 1898. It moved 3200 tons of ore by the latter half of 1899. While the mine earned $56,000 dollars for that yield of tonnage, development costs were high, in addition to horrendous operational expenses. The high cost of fire wood haulage for the mill's boilers effected the closure of the Porto Rico mine in 1899. After a four-year hiatus, the mine started up again for several months, till it closed again. Brief interest was shown again in 1914 and in 1936 after the pegging of gold. The Depression era *Vancouver Sun* prospectus did not expound upon the condition, or presence of earlier machinery. Byron Riblet's machinery was becoming known in BC. In turn, contracts came in from the BC Coast and the US. However, a discussion of the workings of mining tramways is first needed.

Notes —2

1. Don Blake. *Valley of the Ghosts*. Sandhill Publishing, Kelowna, 1988. p. 7.

2. Ibid. p. 65.

3. See footnote 40 in Notes 1.

4. Dave Wilkie and Robert Turner. *The Skyline Limited*. Sono Nis, Victoria, 1994. pp. 27, 46.

5. With Slocan being the nexus and birthplace for Riblet tramways, one would think that there would be an overwhelming volume of information on the topic. In fact, the reverse is true-there is more information on the Conrad, Stewart, Lardeau or Ymir installations. Of the dozen or so Slocan tramways, the author has only positively identified the lineage of half that number. The historic record is patchy to say the least; the *Report of the Minister of Mines* often omits to mention the manufacturer, while the local histories only refer to the trams in passing. With that in mind, the record of tramways in the Slocan will be cited, it will be left to the reader to infer if it is of Riblet origin. On the other hand, Angus Davis suggests most Sandon and Silverton tramways were Riblet products but he does not expressly say it. (See Graham and Davis. *Kootenay Yesterdays*.)

6. Ibid. p. 73.

7. *Report of the Minister of Mines*. 1893, 1895. As quoted in Blake. Op. Cit. p. 74.

8. The original tram at the Noble V was a Finlayson design of Denver, Colorado. It was a pioneer system, and were rarely built outside of Colorado after the turn of the century. (See Viktoria Pellowski. *Silver Lead and Hell*. p. 103.)

9. There is some confusion over the date that this happened. Some claim it was 1896 and others state that it was 1897. The Riblet Tramway Company even notes the discrepancy.

10. Ibid. pp. 105-106.

11. Graham and Davis. Kootenay Yesterdays. Op. Cit. p. 43.

12. These alarmingly heavy, overbuilt engines were very popular in the BC mining camps. Prior to the omniscient and omnipotent presence of BC Hydro, the small BC hamlets were left to their own devices to provide power. Johnny Harris' still extant, prize winning hydro-plant at Sandon is

a good example of this. Other mining camps first used steam, and then later diesel engines. To this end hugh Corliss and Cornish Engines were installed all over the West. The mines of Nevada, Montana, Idaho, BC, Yukon and Alaska are scattered with their relics. Fuel for the concentrator and power boilers was at a premium. In Nevada, the mines burned sage brush. In BC, the Crows Nest coal deposits were developed to fuel the fires. By the First World War, the proto-diesel made it debut. They were heavy, 2 cycle engines with multiple, cylinders and looked like marine engines. The intake cycle scavenged fuel through the crank case, and exhausted, often, unmuffled. These Fairbanks-Morse "Y" engines had widespread use: Stewart, Zeballos, and Sandon all having examples.

[13] "Last Chance" BC Ministry of Mines, MINFILE. www.em.bc.gov.ca/ef/minfile. (A digitized-database synopsis culled from the bound *Report of the BC Minister of Mines*.)

[14] Apparently, the ore freight was split between the two rail carriers. The CPR and GN encouraged the mines to use their facilities even if a mine did not have direct trackage to their concentrator or ore bin. Thus wagon roads were built with the railroad's help. Presumably, this aerial tram was built for the same purpose over to the CPR trackage, even though the Payne mine had ore bins directly over the Kaslo and Slocan Railway. Turner. Op. Cit. p. 114.

[15] Graham and Davis. *Kootenay Yesterdays*. Alexander Nichols Press. Vancouver, BC 1976. p. 156.

[16] *Report of the Minister of Mines*. 1898. p. 974.

[17] *Report of the Minister of Mines*. 1904. p. G190.

[18] Ibid. p. G190.

[19] Ms. MacMillan was introduced to mining through her husband who had properties in the Ontario northcountry in the 1920's. She then began to prospect and develop her own properties. In short, she soon learned the business. She became a successful promoter with the Queen Bess-Viola Mac mine being but one of her properties. Things went well until 1964 when she was charged with insider trading in the Windfall Affair. After that affair, she then glided into grand old age. The affair had large effects as the Toronto Stock Exchange rules were improved and many 'fast' operators moved to the Vancouver Stock Exchange. There, they continued their peculiar dealings and, in course, destroyed that exchange's reputation which eventually forced it into extinction.

[20] Blake. Op. Cit. p. 86.

[21] *Report of the Minister of Mines.* 1904. p. G193.

[22] A spitzkasten was a device used to separate fine materials from the coarse concentrates and not bind up the Wifley specific gravity tables with very fine materials. Ibid. p. G 193.

[23] Clara Graham and Angus Davis. *Kootenay Yesterdays.* Alexander Nicholls Press. Vancouver. 1976. p. 160.

[24] Don Blake. *Valley of the Ghosts.* p. 50.

[25] Royal Riblet Collection. Cheney Cowles Museum Accession L94-50.4.

[26] Blake. Op. Cit. p. 55.

[27] *Report of the Minister of Mines.* 1898. p. 1191.

[28] For details on the Wakefield Mine. (See: Report, 1912, pp. 149, 299.) Robert Trennert implies in *Riding the High Wire –Aerial Mine Tramways in the West* that the trams in the Silverton were built by Riblet. (See Trennert p. 87; Trennert cites Norris' fine book–*Old Siverton*–for support.) This author has not found corroborating evidence to support Trennert's claim, neither in the *Report to the Minister of Mines* nor in Norris' work, though the lack of evidence does not indicate a negative answer.

[29] *Report of the Minister of Mines.* 1907. pp. 99, 219.

[30] *Report of the Minister of Mines*, from MINFILE (www.em.gov.bc/cf/minfile for 'Carnation' or 082FNWO48)

[31] Ibid. p. 1.

[32] Silverton is just south of New Denver. John Norris. *Old Silverton*. p. 144.

[33] Of course the OBU created lasting fame with the General Strike in Winnipeg in 1919. Tensions were running high around the world as the working classes were feeling overwrought by recent demobilization, war deaths, unemployment, inflation, war-profiteers, and events in Russia, Poland, Hungary, Germany, Chehalis, Everett, and Seattle. The OBU was an offshoot of the Industrial Workers of the World and another attempt at a single, large union.

[34] In the Second World War, New Denver became an internment camp for Japanese . During that war's aftermath, Group Captain L. Cheshire VC, of Dambusters fame, attempted to find peace and land there for a returned servicemen's colony.

35 See D.M. Wilson. "Moyie". www.virtual-crowsnest.ca

36 Ibid. p. 2.

37 The Foleys went on to form Foley, Welch and Stewart; another large, contracting firm. It was to become a huge player in the BC forest industry. Later, this concern was modified and later absorbed into Canada's largest forest company-MacMillan Bloedel.

38 Ibid. p. 3.

39 *Report of the Minister of Mines*. 1902. p. 792.

40 Norse mythology makes for cheery reading. Dave Northcross. *Nelson Daily News*. February 18, 1980.

41 *Report of the Minister of Mines*. 1901. p. 838.

42 Ibid. p. 840.

43 George Murray. "A Brief History of Mining in Ymir" *Quartz Creek Miner*. August 1, 1998.

44 See the *Report of the Minister of Mines*. 1901, p. 848, for both the Tamarack and Highland Mines references.

45 Dave Norcross. "Mining History of Ymir." *Nelson Daily News*. Feb. 21, 1980. Reprinted in the *Quartz Creek Miner*. A paper for Ymir's 101st Reunion. Dave Norcross's father was the tram operator in 1904.

46 Ibid.

47 Ibid.

48 The Riblet Tramway Company is not specifically named as the trambuilder, though given its history and activity in the region it is most probable that this was another of its products. Irritatingly, there are times when the *Report of the Minister of Mines*, and local histories all omit to mention the builder of the tram. Wright. "Ymir Yankee Girl Mines" Op. Cit.

49 Article and Advertisement on the Porto Rico and Bridge River mines in the *Vancouver Sun*, 1936. Original is in the Salmo Library and Archives.

50 Ibid. Alas there are few other details on the tram.

Technical Details and the Trenton Iron Works

13. Technical Details Of Mining Aerial Trams

Most modern readers are able to envision aerial trams in the form of the ski lift, though there are differences between the two systems of the mining and ski links. All the same, the chairlift will provide a good basis from which to start the description. Most trams have basic components that are similar: the upper and lower terminal stations, support towers, cable, carriers, tower sheaves, and bullwheels. There are also systems to hold down the towers or stations that are under load, as well as methods to adjust the cable for tension. Similarly, there are braking, drive machinery systems and methods to load and unload. There are different configurations to mining cable systems. The most simple form, and one that resembles modern ski lifts, is called a monocable design.

Monocable

The easiest aerial trams to visualize are the monocable designs. These are the systems of Hodgson, Hallidie, and Ropeways mining systems and all chairlifts. They are a simple loop of cable which provides both power and support for the load. Carriers are attached to the cable and they travel in an endless circle. Monocable systems are limited by load, distance travelled and the steepness of the territory. To overcome these limitations the bicable method was devised.

Bicable

As more tramways came into use it became apparent that the monocable design was not fully able to carry the mining loads. Mining ore is very dense and thus caused much deflection and derailments for the monocable tramways at towers. Deflection is the bane of an engineer's life; it is the behaviour of stretched cable when it is under load. Deflection is the angle that the cable forms with a tower when the cable is under load, thus the tension on the cable and tower is related geometrically to deflection. Too much deflection and a cable system will be damaged.

To overcome these problems the bicable design was devised. It consists of two cables: one stationary track cable that supports the load (typically anywhere from 1 inch to $1\,^5/_8$ inch), and a separate traction rope (anywhere from 1/2 inch to 1 inch) which drew the carriers. Interestingly, the track ropes could be of different sizes dependent upon if it was on the loaded side (down), or the unloaded size (up) that the ore buckets travelled on. Most mining trams were hauling ore down a mountain so the down side needed to be more robust. The track cables were also stationary and needed to be anchored within the terminal and the load force transferred to rock anchors. The track ropes needed a tension adjustment to account for thermal, wear expansion and for heavy load periods.

In multi-section trams, mid-point anchor and tension stations were installed to maintain the proper deflection on each of the support towers. If the system was a 20 mile tramway there might be 4 or 5 intermediate stations—these could also serve as angle stations whereby the tram could change direction, for example to benefit from better ground.

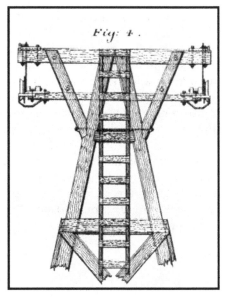

Tower for a Riblet tram. 1903.

The traction ropes were smaller, moving ropes that provided the power to the buckets. Each bucket had a grip that would grab the traction rope as it left the upper terminal—and the bucket was pulled along by the constantly moving traction rope. The grip allowed for intermittent power allowing the buckets to be attached and detached at the upper and lower stations. In this way the buckets were not moving while they were being loaded and unloaded. It was a much safer system. Many bicable operations were what was called the Continuous Type. All the buckets traveled in unison first over the up side, then after reversing direction in the terminal station, returning to the down side.

At the upper terminal, the timing of the buckets was controlled to limit the number of buckets between towers. Only so much deflection of the track cable would be within the margin of safety. To space the buckets, the operator had a bell or timer to denote when a bucket was to be despatched. All he had to do was push the bucket, and through an inclined start up ramp, the bucket picked up speed and left the ramp to be transferred to the track cable. Usually the bucket grip was held open within the station—to allow free movement—but as the bucket left the station a toggle was switched which applied the grip. By the operator's simple push, the bucket was accelerated to operating speed, moving on the track cable, and activated the grip on the traction rope—a process which only took several seconds until the bucket was on its way.

At the bottom terminal the process was reversed. The bucket came into the station and it was switched off the track cable to a slightly raised steel frame (which slowed its movement), while a metal jig reversed the grip toggle to the release position. Thence the grip would let the traction rope go. The buckets would be manually emptied (they pivoted like a mine car) into the ore bins then moved through the station and back to the mine (often the buckets were inverted to prevent the buildup of snow or rainwater).

Automatic unloaders were common too: lugs on the ore buckets allowed metal activators in the lower station to lift the lug and bucket to the point of overbalance, and the bucket would tip into the ore bin. Such stations were unmanned.

The Stations

Stations were an agglomeration of mine buildings. The top terminal was amidst bunkhouses, cookhouses and workshops. Ore was taken from the mine by ore car to the storage bin. Huge wooden hoppers were intricately constructed with 10 x 10 inch wooden posts, all bolted and rabbetted into a large frame. Panelling the frame was stout 3 x 10 inch planking. Slightly below the bin was the upper terminal. These were equally stout structures as they needed to support and transfer the load of the track cables, provide a foundation for the traction cable grip wheel, sheaves and bucket turn around track steel, brakes and other drive mechanisms. The ore bin allowed for gravity feed into the tram carriers. These bins and top loading stations were roofed wooden buildings needed to keep out the weather and give some protection to the machinery. As the terminals were open at one end—to allow the cables and carriers to enter—the terminals were cold and subject to the wind. Woodstoves were sometimes used to heat such spaces.

The Towers

Wooden, pyramidal towers were perfected to carry the cables. These towers look much like modern high-tension electric towers so common in late-modern BC. Having four faces, these aerial tram towers were the most stable and sometimes rose to a hundred feet though usually most were twenty or thirty feet in height. The track ropes were carried in a cast saddle, which spread the heavy load over a wider section of the cable to prevent the internal wires from breaking (which is fatal for a cable). Sometimes the saddles were pivoted to allow them to move with the varying deflection of the track cable. The towers had a T-section at the top, which extended the saddles and traction rope sheaves to allow the buckets to clear them. Below the track rope was the traction rope, and it was supported on a sheave. Sheaves could be narrow or wide depending upon how much lateral movement there was on the traction rope. Sometimes there would be hoops to prevent the traction rope from slipping off the sheave. Depending upon the load, the buckets would cause oscillations in the cable and thus shake the tower. The tower bolts needed to be inspected for tension periodically.

The foundations of the tower varied by operation. Some later ones had cement footings, while the early towers simply had the four tower feet on the ground. Other installations used a wood crib to form the base of the tower. Obviously, the unsecured towers could move out of alignment so a solid base was important.

If the cables came over a pronounced hill—one where the cables formed a significant arc–more than one tower would be needed. This was called a breakover station. Again heavy deflections on the cable had to be engineered against. "A curved rail is fastened just above the cables for the carriers to move on. This prevents excessive wear on the cables and provides a smooth circular track for the load to move on instead of a series of short chords."[1] Sometimes if the cables pass over a long unsupported span,

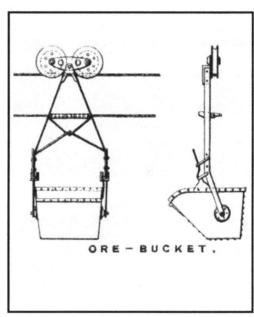

Riblet buckets were angular in shape for loading.

a breakunder section is needed (the cable travels under the sheaves at the tower) to maintain tension on the cable. Too long an unsupported section allowed for slack in the cable which could allow the traction ropes to slip from their sheaves.

Tall towers sometimes were guyed to secure them more firmly. Bleichert used very tall towers with small bases, and thus found guying necessary. Later, the tram builders stopped using wood towers and substituted steel for durability. Towers also came in a wide variety with pyramids, vertical lattice, and through frame varieties. Riblet primarily used wooden pyramid towers and later switched to steel for remote installations. Steel was easier to transport.

There are several good technical books and articles on the design of mining tramways. Peele's *Handbook of Mining Engineering* is one and Metzger's "Aerial Tramways in the Metal Mining Industry" is essential too; both are math based and calculate angles, loads and deflections. Schneigert's *Aerial Tramways and Funicular Railways* is detailed, though from a modern, Middle European perspective. [2] It also does not mention any New World operations; probably due to it being Cold War era and within the Eastern Bloc at the time of publication.

The interior of a Trenton Iron Works tramway at Granite, Montana. c.1893.

Applications

Aerial tramways are ideal in steep, mountainous country. They are also good for places were one must cross large rivers or ravines. Moreover, aerial tramways have proved themselves reliable in locations of high snow or inclement weather. As the trams travel above the snowpack, they are unhindered by winter. The sheaves and cables need a minimum of de-icing to prevent cable derailment. As stated tramways are not limited to only mining ore: lumber, sugar, bananas, linoleum, molasses, tea, sand, gravel, earth fill, coal, passengers or wheat can all be hauled.

Carriers

In most of the mining applications the tramways used cubic steel boxes to shift the ore. Some manufacturers have used cylindrical bins instead of the square shapes. To and fro carriers used much larger carriers to increase capacity while using a fewer number of individual carriers. Of course, tramways were versatile and able to adapt to whatever material needed to be hauled. Timber hooks—or flat platform—were developed to move pit props. Individual carriers were also developed for bricks, molasses, bananas, textiles, or boxes for example. The size and capacity was only limited by the engineering of the actual tramway.

On bicable tramways the carrier used a truck, with wheels or sheaves, to run on the track cable. Sometimes, the sheaves had changable steel insert tires, as modern lifts have rubber liners, to offset service wear. While the truck acted as a mount, securing the sheaves with bearings, it also provided connection for the dropped pendant bracket. The pendant bracket or hanger could suspend from the sheaves and adjust in angle according with the terrain. The bracket also provided a stout frame and a place to mount the traction cable grip.

Finally, slung below on the steel frame was the cubic ore bucket. It rotated on lugs which facilitated loading and unloading. Latches, hasps and automatic unloading mechanisms were also sometimes added depending on the need. With this combined mass, each bucket and mount weighed hundreds of pounds when empty.

Grips

These were a key device of aerial tramways which allowed the carrier to grab the traction rope. Grips needed to grab and release at the terminals and they also were required to be very reliable in all conditions. While the individual designers experimented and came up with variations, they fell into two groups of types.

Fixed Grip

This was where the carrier was firmly and permanently bolted by a bracket to the cable. It could not release as the bracket was often woven into the cable.

Detachable Grip

A detachable grip used a mechanical clasp and allowed the buckets to grab and release while at the terminals. Bicable tramway designs had separate carrying ropes and traction ropes and were designed for the clasping and releasing of the traction cable. This was done with either brass teeth or a bronze bearing plates that were opened or closed by the grip. Bleichert, Trenton and Leschen used weighted levers, which applied pressure to the traction cable via a screw and bearing plate (the Webber and Wico grip). The position of the lever precluded which state the grip was in.

Screws were also used to affix grips (the Otto Grip). Riblet used springs and wedge activated grips to clasp the traction cables.[3] It must be said that all companies experimented, designed, patented and improved their grips over the years. In sum, there is no easy formula for identifying which grip was used by which company or tram installation as changes and substitutions were common.

Riblet had several patents for bucket grips—they involved the use of springs and wedges. By the 1920's the spring Riblet clip was perfected and it used a small activator lever to automatically open and close the grips. Modern high-speed chairlifts have similar grips on their arms.

Jigback or Reversible Tramways

For smaller loads jig back tramways were built. Also called reversible or to and fro tramways, these filled a need for applications of less than 2000 feet and under 50 tons capacity of ore per day. Many of the smaller mines built them. Modern passenger gondolas work on the same principle as the carrier reverses operation when at station. As with the larger bicable installations, the jig back used track ropes and traction ropes. Where they differed was that the jig back trams only used two carriers. In many respects the jig back resembled the modern gondola or funicular railway in that when one carrier was ascending while one was descending.

The smaller installations allowed for tinier terminals and ore bins, which reduced the costs. Riblet had a complete line of jig back tramways for smaller mines. They could be built to eight foot gauge or four foot for a diminutive operation. Of course, the simplest tramway would be a single-track rope and one traction rope moving a single carrier back and forth.

Riblet Jig-Back tramway from 1920's.

14. The Competition

Adolf Bleichert and Company

The Riblet Aerial Tramway Company was neither the first company to built industrial mining tramways, nor was it the most prolific builder. That honour would go to the Bleichert Company of Leipzig. Bleichert would become the world leader, develop influential patents, and lead the way in mining material haulage. The advent of the Bessemer Converter—and its output of cheap, quality steel—allowed industry to expand the scale and size of machinery in the last quarter of the nineteenth century. As with other inventions and developments of this era, the aerial tramway flourished and became workable in this period chiefly due to the availability of steel cable.

Theodore Otto and Adolf Bleichert tinkered with tramways and built one to carry coal in the Ruhr district in 1874.[4] It was a moderately successful monocable design. In the ensuing period these German engineers experimented and improved an aerial tramway design much in the manner that Hallidie was doing in San Francisco. Otto split from Bleichert and set up his own firm in 1876.[5] Meanwhile, Bleichert perfected the first bicable design of tramway. That is, unlike the monocable, his systemwould have a supporting, stationary track cable, and another power or traction cable to move the bucket.

In America, the Cooper Hewitt Company bought the North American patent rights and proceeded to construct trams using the German designs. By 1890 Cooper Hewett and its manufacturing arm Trenton Iron Works was a major competitor of Andrew Hallidie and his tramways. As the bicable model proved its worth in the unforgiving terrain of Colorado, Trenton began to triumph in the industry. A 1892 pamphlet by Cooper Hewitt extols the virtues of the Trenton system and describes some of the Rocky Mountain installations. One at Creede, Colorado was one of the steepest installations anywhere: it ascended an astonishing 1 in 3 from the canyon floor to the mine adit.

The Trenton-Bleichert systems were installed, as stated, at the Chilkoot Pass, two trams at Kennecott, the mines on Prince of Wales Island, (on the Alaska Panhandle;) North Star, Jumbo, and Dorotha Morton mines in BC; at Monte Cristo and Winthrop in Washington state; at Mackay, Yellowpine, and Kellogg, Idaho; Cornucopia, Oregon;[6] Shasta, California; Sopris, Alta and a dozen other mines in Colorado; Bingham, Utah; and gold mines in the Philippines.[7] Trams were also built in Mexico, in the Eastern US for coal, sand, and materials haulage. They also built a lumber carrier in Haiti to log the mahogany.[7] Another large lumber carrier was built in Northern California at Spanish Peak.

The full extent of the parent Bleichert activities in Europe, England, Russia, South America and (through its patent licensee Trenton Iron Works) North America are far too numerous to mention in other than a cursory manner. In fact, even a book would hardly do the topic justice. Bleichert dominated the industry for sixty years That said, there is one tramway that encapsulates the sweep and market control Bleichert had and that would be the large tram at Chilecito, Argentina.

The silver mines of the interior of Argentina were located in the high Cordillera near Cordoba. To transport ore the German Bleichert Company was contracted to build a 22 mile long interlinked tram. Its loading station was at 14,000 feet making it one of the highest trams in the world. The tram descended 11,000 vertical feet in that distance. But because of the undulating nature of the ground, the tram was built in eight sections. Steam engines were installed to smoothly move the ore buckets on the horizontal, while braking and governing systems controlled other, steeper portions of the system.

An early Bleichert tramway at Halle, Germany in the 1870's. It moved coal in the Ruhr mining district.

All the material was packed in with mules from the railhead. The elevations and privations of the terrain involved made it an overwhelming construction project. In the tramway world, Chilecito is a special tram and nearly a century on—it is still spoken of. The tram was built in 1907 and ran for decades. The steel towers, tension and terminal stations—combined with masonry buildings—added a permanency to the operation. Interestingly, the tram is still exists in working condition no less—one of the few trams in the world to have been preserved. As a result the Chilecito tram has become something of a tourist attraction, and the district was recently giving tram rides in the buckets over a portion of the towers. Sadly, a tourist fell from bucket in the 1990's and the practice has since been discontinued.

Bleichert of Germany also built trams in Poland, Hungary and Czechoslovakia. They also were on contract to build mining cableways at Russian asbestos mines during the period of illicit Russo-German military trade and its concomitant commercial trade in the Weimar era.[9] Consequently, with the German tramways as patterns, the Soviets went on to design their own systems as they so often did when they copied other durable Western products such as the Model T Ford, Caterpillar 60 tractor and Douglas DC3 transport plane.

Bleichert expanded its market trying electric cars and later going into large earth moving machinery. In Germany, large continuous cutter head and conveyor machinery is used to strip lignite for thermal electric generation. Bleichert relocated its factories into Western Germany after the Second World War and is still in business making material handling systems for industrial factories. A branch of the company merged with German competitors to form PHB, then the largest trambuilder in the 1980's. Subsequently, all the European companies have been merged into the Dopplemayr concern. In America, the Trenton Iron Works was bought out by J.P. Morgan's United States Steel and became a division of US Steel's American Steel and Wire Company just prior to the First World War. It continued to build tramways until they became obsolete in the 1950's.

The Broderick and Bascom Company

These were cable manufacturers from Saint Louis who were in competition with A. Leschen and Sons. Broderick and Bascom first supplied cable for cable cars and for general purposes. They later built tramways in their own right. Broderick and Bascom succeeded with the Hercules line of cable. It was durable enough that it won the contract to supply the cable for the large Lidgerwood cableways that allowed the massive steel Gatun Locks to be built in the Panama Canal. Later, the Interstate Company bought the tramway division out and built cableways in their own name. Broderick and Bascom are still in business today selling wire rope, though details of their aerial tramways are very scarce. However, the discussion has wandered far afield, and we shall return to Byron Riblet and his actions around BC.

15. The A. Leschen and Sons Aerial Tramway Company.

The A. Leschen and Sons Company played an intriguing role in the Riblet story. First it was a competitor, then a parent company and employer to Riblet, and later a partner and parts supplier. Untangling the connections between the Riblet Tramway Company and the Leschen Tramway Company is difficult, if not impossible.

The A. Leschen Company of Saint Louis started out in 1857 as a general steel cable manufacturer. It did well with the expanding west and extended its market into cable for street cable cars. As said, it soon added mining cableway systems to its market line as well. One of the larger Leschen tramways was the large Silver King tram at Park City, Utah and was finished in 1901.[10] Leschen would later go on to build many systems in the West: famously at Cerro Gordo, California, Encampment, and around Mexico. The tram at Pioche, Nevada carries buckets labelled with the Leschen name.

Byron Riblet sold use of his grip and loader patents to the Leschen Company in 1903. He then was installed as Chief Engineer for the company and excelled with the Encampment Tram in Wyoming. It is claimed that the Britannia Mines tram, Encampment tram, and Sheep Creek tram were Leschen products, even though Byron Riblet played a part in their creation. In the Edwardian era, Leschen claims to have built "100 trams" around the West.

The contract details between Byron Riblet and the Leschen Company were as follows "(Byron Riblet) to get 6% of gross business of Leschen. $300,000 guaranteed."[11] The Leschen contract went on to stipulate what Royal and Walter would receive. "W.S Riblet and R. N. Riblet to get from B.C. Riblet 1/6 of 6% of gross business of Leschen. R. N Riblet and WSR to divide the 1/6 equally. Salaries of RNR $200 per month, WSR $175 per month."[11]

Many Leschen tramways were built in the Eastern Coalfields and at gravel operations. One Leschen system for a Western gravel operator was at the Ross Island Sand and Gravel Company in Portland, Oregon. It hauled gravel from a borrow at Ross Island—in the Willamette River—up to the loading plant and railroad on the East bank.

Leschen did maintain a healthy parts business for aerial tramways. Obviously, they were supplying the larger machine parts and castings for BC tramways. Two grip bullwheels from their factory survive and are on display at Riblet installed tramway sites: one is at Silverton, BC from the Standard Silver Mine, and the other is at Coalmont. The mixing of parts and engineers seemed to be common in this period.

Byron Riblet broke from Leschen in 1908 when he resuscitated the Riblet Tramway Company in Spokane. Leschen continued with the tramway business until 1953, though was severely affected by the Depression. It is still in the wire rope business in St. Louis to this day.

Notes—3

1. Metzger. "Aerial Tramways in the Metal Mining Industries". *USGS*. Washington. 1937. p. 9.

2. It appears that there is mystery and intrigue behind this publisher. Pergamon Press, it seems, was established in 1947 by a Czech *émigré*. His desire was to translate and publish good, technical manuals into English, and so a Polish work on aerial tramways appeared. In the process the publisher Anglicized his name to Robert Maxwell. The details of Maxwell's dealings and financing are murky. Some suspect he was a KGB spy sent to collect published materials in the West which may explain his Eastern Bloc contacts and ability to publish fine works. Maxwell went on to form a Press Syndicate with several English newspapers in his bag. Scandal and mystery followed him to his death. While yachting on the Atlantic, he strangely fell over the taffrail of his boat and drowned. This is after he looted the pension funds of his employees to bolster his failing press empire-some suspect murder by darker elements in society.

3. Robert Peele. *Mining Engineers Handbook*. John Wiley. New York. 1941. Third Ed. Vol. II. pp. 18-26.

4. Zniebnow Schneigert. *Aerial Tramways and Funicular Railways*. p. 6.

5. Trennert. *Riding the High Wire*. p. 30.

6. The gold mines at Cornucopia, Oregon are in the beautiful Wallowa Mountains. Their Trenton system became operational in 1916, unfortunately, there were difficulties with it-high winds forced the the cables together which caused a derailment. Apparently, the Riblet Tramway Company had a contract to do work on this tram, though the date and nature of the work are not known.

7. During the American Army recapture of the Philippines in 1945, the US were forced to fight street by street in Manila and thus the city was mostly demolished. General Yamashita, the Japanese Army commander during the hopeless and pointless defence of the Philippines retreated the mountains and a more defensible position. Thus he wound up in the gold mining town of Baguio.There he held out against the superior American force despite the lack of food and supplies. The US used its air superiority with devastating effect and the mountain city was reduced to ruins as well. To escape the air marauders the Japanese found available shelter:

> **Most of the hospital patients had been taken to a mine, the shaft of which was two miles long. At one time there were six hundred**

people in this mine. Nurses and orderlies used ore cars to move patients and bandages along the galleries. This improvised underground hospital was damp and airless, and water poured from the ceiling. Hundreds of badly wounded soldiers lay moaning on the floor, while nurses groped around in semi-darkness slipping on blood and trying to look after the patients as best they could. The smell and the noise were horrible. Some men raved in delirium others kept groaning.

From: John Potter. *The Life and Death of a Japanese General.* p. 33. Yamashita, in this way, was able to hold out against the superior forces until the final Japanese surrender by the Japanese Government in August, 1945. The Japanese also inflicted suffering in their mines. Chinese, Korean nationals and Allied servicemen were forced to work as slave labourers in China, and the Japanese Home Islands.

The nearby Philippino gold mine of Benguet had a Bleichert tram. It was also used to move pine and cedar timber that was found in the mountains down to market and it was running in the 1950's. The Benguet Company later moved its odious logging to BC-they removed the cypress trees at Cypress Bowl park. Taiwan had a copper mine, aerial tram and concentrator on its northeast shore. This operation, along with other mines in Korea, were extensively bombed by the US Army Air force in their island hopping campaign.

[8] Logging in Haiti is contentious. Haiti is now 98% deforested. Commercial logging, charcoal burning and subsistence farming are the main culprits. Outside investment and involvement are some of the root causes-a large population is another reason. The US occupied Haiti between 1915 and 1934 to protect its citizens and investments-the intervention was purportedly due to the passing a law preventing foreign ownership of land. Afterwards, under the Duvalier regimes, complete degradation of the country ensued.

[9] See Smith. *Joe Hill-the Man and the Myth.*

[10] Sadly few other details regarding the connection of Byron Riblet and Leschen and Sons have surfaced. Royal Riblet Diary. 1903. Cheney Cowles Museum. Spokane.

[11] Ibid.

The Lardeau and the Kootenay Lake Mining Districts

16. The Lardeau District, B.C.

Lanark Mine

The Lanark Mine was a very early mine and tram expansion in BC. Located adjacent to the Canadian Pacific Railway mainline on the Illecillewaet River, part of this operation used a 3000 foot Otto aerial tram to access the mine from the valley floor. It stretched from the valley floor in a single 1500 foot arc to a tower atop a high cliff. The Selkirk Mining and Smelting Company installed it and they also constructed a hotel, mill and manager's building on the flats by a railway siding. This was not far from the newly completed Canadian Pacific Railway tunnel at Illecillewaet. The Lanark mine was staked in 1883 with production starting in earnest in 1888; the vein being worked for five years. After which time the mine became dormant.[1]

Around 1907 another ore shoot was discovered in the old workings and the mine and plant was rehabilitated. This included the Riblet Tramway Company replacing the Otto tram with a Riblet one. The new tram was built in two sections to create a tram 6900 feet in length.[2] The Lanark mine had a new lease on life for ten years and produced profitable galena ore.

Ferguson

When the Canadian Pacific Railway was completed across Canada, with the last spike being driven at Craigellachie in 1885, workers dissipated across the land. Some looked for minerals. The discovery of silver on Toad Mountain above Nelson illustrated the richness of the Kootenay country and, correspondingly, prospectors looked in the nearby mountain regions as well.

South of Revelstoke, lies the Arrow Lakes; they form an easy communication link to the south by canoe. By passage up Kootenay Lake, Lardeau River, and Trout Lake—from Nelson—one can complete a round trip. Andrew Abrahamsen started such a trip on foot in 1890 from Revelstoke and walked with his dog and cat to Trout Lake. There he built a cabin on the lakeflat and staked the Maple Leaf claim.[3]

Mere words cannot convey description of the Lardeau territory. It is a beautiful, forbidding, and unforgiving area. It is very rugged with huge cedar and hemlock forests, swamps, interspersed with the perfidious plant Devils Club. Mountains cascade up to eight thousand feet with steep slopes and rockfalls. The weather is not forgiving

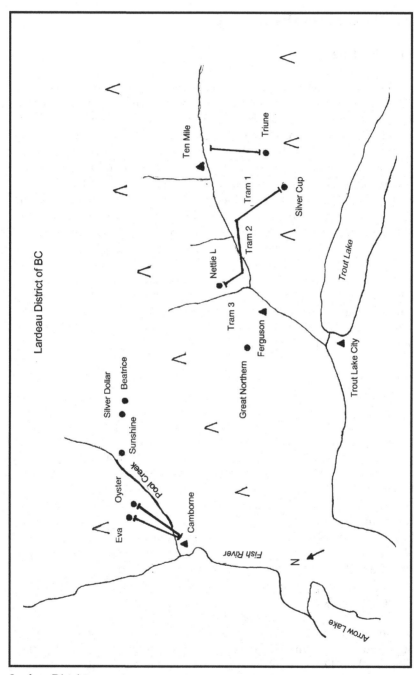

Lardeau District

either: it is a wet territory as it is on the weather side of the Selkirk Mountains, and in the winter that rain falls as snow. The snow and ice stay for months. Danger and peril abound, including grizzly bears.

As accessible mineralization occurs at high elevations, the poor prospectors had to climb though dense forest, camp, and explore the rock outcroppings, though sometimes they lit the forests alight to watch the smoke colour, and mineral content, while the trees were consumed in flames. Ore discoveries are made where rocks are exposed and these are usually the eroded peaks high in the mountains.

Humans continued with their courting of the country. "Not all who came were after gold and silver. J.W. Thompson, a school teacher from Revelstoke, had already bought land at the head of the (Arrow) Arm—which he called his ranch—and was intent upon eventually subdividing it to establish a townsite."[4] Others—including David Ferguson—came to prospect, and they wintered to improve and develop their claims.

A huge vein of galena was located and it contained valuable gold and silver. This vein ran in a northwestern direction and was exposed in numerous locations: at the South Fork of the Lardeau River, at the confluence of the North and South fork (near the future town of Ferguson) overlooked by Great Northern Mountain, and the north side of the mountain accessed by the Fish (or Incommapleaux) River.

Thus interest continued in the region. Not a rush or stampede as in Sandon or the Klondike, yet steady progress by 1892. By that year steamboats plied the Arrow Lakes making connection easier. Trout Lake and Thompson were platted and houselots sold. Work commenced on wagon roads to Trout Lake and the Fish River. A severe winter washed out bridges and roads in 1894.

Nonetheless, the miners toiled extracting the rich ore, sacking it in rawhide bags and hauling it by horse and sleigh over the frozen ground to the lake. Yields of ore were in the 500 ounces of silver per ton range with that shiny metal selling for around sixty cents an ounce. Despite the returns, the development work was onerous. Roads and pack trails were built, as were cabins. Supplies had to be loaded and freighted to staging points. Finally the supplies hauled up the steep grades to the claims, while the ore rawhided out. Mines such as the Badshot, Silvercup, Great Northern and the Nettie L became big producers; while hundreds of other claims were registered as well in this period. Each series of claims needed its supply point and with the claims being separated by mountains and rivers, different towns sprang up.

Thompson's Landing was the steamer landing on the North Arm of Arrow Lake while Trout Lake (sometimes embellished as City) was superbly situated inland by the mountain lake. Both camps prospered. Nearer the claims, some five miles up the Lardeau River was another flat region and the townsite there was named after its promoter of David Ferguson. Money was to be made speculating in lots, peddling rooms, food and drinks to miners, and waiting for the boom caused by the arrival of the railway.

The Lardeau Region (or sometimes Lardo) thus was on the map. Hotels were completed in Trout Lake—these included the Windsor and the Trout Lake. Stores and mining supply retailers were also established; as was a newspaper, church and school. A steam sawmill was installed on the Trout Lake to take advantage of the abundant wood and supply the mines, and supply townsites with lumber.

Rivalry increased between the two towns of Trout Lake and Ferguson. As they were close together, competition for business, fame, and boosterism was intense. This rivalry played out on the baseball field, duelling newspapers, and jockeying for position with freight companies.

New steamers plied Arrow and Trout Lakes: the *Lytton, Marion* and *Nakusp*. Ore was shipped to John Rockefeller's smelter in Tacoma via the Arrow Lakes and the Great Northern Railroad or the CPR. Augustus Heinze would soon complete his smelter in Trail. The country was being opened up. Yet all was not progress, one's lifespan was tenuous in this dangerous territory. Men were dying from dynamite blasts, in snowslides, failed rivercrossings, and a ship sinking. Yet 1897 "had been a momentous year in the Lardeau. Tremendous growth and promising ore finds kept most settlers and investors confident there was good fortune ahead."[5]

An English syndicate called Horne-Payne bought out the productive mining claims from their original prospectors. In this way the Silver Cup and the Great Northern came under the syndicate's control. The prospectors earned $106,000 for their efforts.[6] Thus the nature of mining changed from the small player staking his claim and prospering, to control by consortiums. In a region like the Lardeau, consortiums were needed for they had pools of capital from which improvements could be made.

Parm Pettipiece, the colourful editor of the *Ferguson Eagle* newspaper, opined about the state of things "No machinery, no transportation, no capital, no nothing, but mountains of silver ore. That's the Lardeau's condition"[7]

David Ferguson constructed a sawmill and ornate hotel in his eponymously named mountain meadow. The miners continued to dig, and sack at the rich, exposed silver veins. The Eagle expressed the economics of mining: "Each miner is taking out besides the development work, about 700 lbs of clean shipping ore a day, which gives the Great Western Mines company $52.50, while the miner gets $3.50 for his work."[8] Despite the profits, it would take five more years to complete mining development with operating concentrators and tramways.

BC had just passed progressive labour laws regarding the eight hour-day. Wage scales were also negotiated for the tenacious mining union—the Western Federation of Miners—had battled the owners and legislators for both rights and had won. Meanwhile, labour battles were fought in Colorado, Coeur d'Alene, and Rossland. Some of the miners in the Kootenay were blacklisted men from the infamous Bunker Hill mine in Idaho.

Also, other players vied for profits. The Great Northern Railroad and the Canadian Pacific Railway continued their feud. Both railways had built rival lines into the Boundary and Kootenay mining districts. They had even come to blows in Sandon and Midway. Thus: "A new fight is underway between the CPR and the Great Northern for the trade of the Lardo-Duncan districts of BC."[9]

Ruinous competition was a hallmark for business in the nineteenth century in much the same way as telephone companies battle today. Cooperation, network sharing, or judicious expansion of systems was non-existent. The Great Northern's thrust came in the form of a proposed Kaslo and Lardo-Duncan Railway. Its route would start at Argenta—at the north end of Kootenay Lake, and carry on up the Duncan and Lardo rivers to Trout Lake. A charter was issued and the legal and financial details agreed. The projected cost was $ 1,990,000 for construction.[10]

Meanwhile, the CPR planned a extension of the Arrowhead connection at the northern end of Arrow Lakes. It would travel around the North Arm, across the Fish River to Thompson's, and on to Trout Lake. A spur would finally connect to the mines of Ferguson. South of Trout Lake the railway would follow the west bank of the lake and follow the river the forty miles to Kootenay Lake. Surveyors arrived and commenced work in the summer of 1899.

Despite the heated beginning, interest in the region waned. The Great Northern Railway never laid a mile of track, and the CPR only completed a line from Kootenay Lake to the southern shore of Trout Lake. That line was finished in 1902. A link to the mainline at Revelstoke, via Arrowhead, was never started. Connecting steamboats were still required to link up to Kaslo, Nelson or Arrowhead. A steamer also worked the waters of Trout Lake.[11]

Ore travelling from the Silver Cup would have to be handled multiple times to get to Nelson or Trail. It was sacked and rawhided to Ferguson, moved by sleigh or wagon to Trout Lake, then hand loaded onto a steamer, and re-loaded onto a southbound train bound for Kootenay Lake at Gerrard. Freight, people, and mail would have had to follow the same involved process.

A concentrator would reduce the distance raw ore would need to be hauled. By removing the dross, only concentrates would need to be shipped to the smelter, saving on freight rates. Thus by 1900 a concentrator became a priority.

A civil engineer named George Attwood was contracted to build a tramway and concentrator for both the Nettie L and Silver Cup mines. He "devised the engineering scheme...for the transportation of the ore and the extraction of metals on the spot."[12] As the ores contained 24% lead, 11% zinc, 180 ounces per ton silver and 6 of gold, it was very rich.

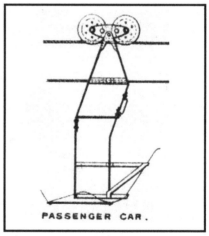

An early chairlift: a 1903 Riblet chair for the Silver Cup tram; Ferguson, BC

The concentrator works were located just east of Ferguson at the junction of the North and South Fork of Lardeau Creek. This camp was soon called Five Mile. The Nettie L mine was nearby, atop mountains to the northwest of the works. The Silver Cup mine was further up the South Fork near Ten Mile (viz-10 miles from Trout Lake in a south, southeast direction) high on a ridge at 6900 feet altitude. In his report Attwood stated "Four aerial trams have been constructed for conveying minerals from the mines the reduction works, and also for conveying mining and food supplies, mining timbers and firewood as well as passengers, up to the mines."[13]

It was an elaborate system with interlinked trams, relaying the ore between elevations and angles. Three tramways were linked to connect the Silver Cup mine to the concentrator. Number 1 tram connected the mine to the valley bottom running 7887 feet long while dropping 2500 feet in that distance. One tension station was needed to prevent the cables from sagging and twenty-three towers were used to carry cables on Number 1's system. (See map detailing tram placement in the Lardeau.) While the Number 2 tram was $3^1/_4$ miles long and falling a respective thousand feet. Starting from the bottom of Number 1 tram, Number 2 emptied into the concentrator. Two long spans, of 2000 feet, were sited on it to cross Lardeau Creek. Two tension stations were also needed to take up the slack from those expansive spans, while forty towers were used to elevate the cables off the ground. Forty-six buckets worked on this tram.[14] Tramway Number 4 was a short feeder tram just over 1000 feet in length and it connected the upper workings of the Silver Cup Mine with the top terminal of tramway Number 1. Five towers raised a 7/8ths inch track cable over its length. Only eight buckets were used on this link. Number 3 tram was an individual link that ran from the Nettie L Mine down Great Northern mountain to the concentrator at Five Mile.

All the trams were built by Byron Riblet—with the help from his brother Royal Riblet. The usual double rope system were employed, one track cable and one traction cable—with the Riblet system of loading, grip, and towers. Counterweights maintained the tension of the one-inch diameter track cables at the lower station. Self-gripping bull wheels returned the 5/8th inch track cables from the heavy to the light sides (up and down). Most mining tramways were gravity powered as the ore had weights running about four pounds per cubic inch. With such a weight of ore in the heavy 'down' side, there was enough energy to overcome friction and the work of raising empty buckets on the 'up' side. If anything there was too much energy and a braking system was needed to retard the traction rope to prevent a runaway—it was a chock on the traction rope grip wheel located at the upper loading station.

An automatic loading hopper loaded pre-set volumes, from the ore bin, into the thousand pound load, steel tram buckets. One tram operator controlled the loader's motion. Twenty buckets were used on the line, for an overall capacity of ten tons per hour. The buckets moved at three miles per hour when the Riblet bucket clips latched onto the travelling cable. As the buckets of one tramway entered the bottom station, they were automatically unloaded into an intermediate ore bin. A second automatic loader charged the Number 2 tram buckets.

As standard equipment for the Silver Cup mining tram, a travelling chair was built. Because Riblet tramways were versatile, various attachments were added as need required—timber hooks for moving pit props and oil drum carriers were common for instance. The 1903

View of Camborne with Eva and Oyster tram lines behind. Oyster mill in background. Photo Vancouver Public Library 107

drawings for the Silver Cup tram include a passenger chair. In effect, this possibly makes the Lardeau District Riblet tramways the first recorded use of a chairlift in the world.[15] Miners and other passengers had traveled on other North American mining tramway buckets before in their day-to-day travels in the mountains but here was a specifically designed chair for travelers.

One reporter for the *Ferguson Eagle* took a ride on the Silver Cup Tramway on the special chair built for the Riblet tram. He:

> climbed into the comfortable carriage built something like a Morris chair and in moment was launched into space, the car being propelled in mid air, sometimes 40 and sometimes 90 feet above the ground, and in a space of 40 minutes was landed on the top of Silver Cup mountain.
>
> The start is the only time one thinks of possible danger, when the car shoots off from the lower terminal across a span of 1800 feet to the first tower. Then 90 feet below rushes the swollen south fork of the Lardeau river looking, from the car, angry in the hot sunshine that the melted the snow of the mountain side so rapidly. One loses all thought of danger as they soar up and over mountains, across gulches and up, upto the white snowy glacial peaks above. Sitting in the car and gazing about, the scene is certainly picturesque, but better even is it on the return voyage, descending from the lofty peaks with the sun setting in the West.
>
> The right of way for the tram is cut about 100 feet wide through an evergreen park-like woods of excellent timber, and the tram buckets pass along just about on a level with the tops of the trees but the view was not impeded with.[16]

This is amost interesting, if not flowery, account of an early chairlift ride. While skiers and tourists today can repeat the experience at many ski resorts, in 1903 such installations were not available to the average public. Once the Silver Cup tram was in operation, all supplies were brought to the mine. This was a great advantage as the mine was in the high alpine region on Silver Cup peak.

At Five Mile, many buildings were erected but a shortage of lumber hindered construction. Large timbers were shipped in from the coast, while the others were sawn in the company's mill at Ferguson.

> The heavy tram cables were being brought to Five Mile by [the packer] Daney, and from there to the site by R. Hardy. Each section of cable weighed from three to seven and a half tons so eight horses were needed to make one trip a day. Between the Silver Cup and the Nettie L companies, the largest part of Ferguson's income relied upon their continued investment.[17]

The Ferguson smelter was started in August of 1902 to great fanfare. "The new Vulcan smelter got a preliminary trial run...."[18] The *Ferguson Eagle* continued "the coke, ore, fluxes run into the furnace. In about an hour and a quarter, the whistle sounded as an intimation that the actual operation of the smelting was in progress."[19] The separated silver metal was removed, and once cooled, shown to the great delight of the populous.

The Five Mile concentrator was a modern plant when installed in 1903. It had a ten stamp mill to pulverise the ore to a powder. The ore was mixed with water into a slurry whereby it would by passed over a spitzkasten to divide the very fine material (which bypassed the next step) from the coarser concentrate. The slurry was then passed onto the Dodd buddles or separating tables which used the difference in specific gravity of the ore contents and an oscillating, stepped cone to separate the minerals. These tables were essentially industrialized gold pans.

The various grades of concentrates would then be dried in a roaster and sent to the furnaces to be mixed with chlorides. The fiery reaction would further drive off sulphur—the element so common in natural state minerals. Other refining was done to separate the gold. After this the concentrates were bagged and sent for final processing at the smelter in Trail, some 150 miles away by boat.

Behind the fanfare, there were mineralogical and technological problems. Apparently, the smelter was insufficient. In turn the plant at Five Mile was upgraded with a chloridising plant in an attempt to obviate mineral separation. The mill ended up treating some 10,000 tons of ore by 1905.

With such progress, work continued on the tramway to the Nettie L mine.

> One hundred and eighty men were employed in the Ferguson mines in October of 1903; ninety six at Five Mile, from twenty to twenty-five at the Nettie L, and from sixty to sixty-five at the Silver Cup. Although the tram derricks from the Nettie L were in full view from Ferguson, there was no possible way for the tram to be finished before winter.... At the Silver Cup, men were building

a short tram which would run from the Upper Workings down to
the upper terminal of the intermediate tram. A large force was also
building a new boarding house and a bunkhouse on a new site out
of reach of possible slides.[20]

With the tram complete, the Nettie L became an important mine and large employer.

Mattie Gunterman, the pioneer photographer, worked for a time as a camp cook at the Nettie L and took some wonderful, human photographs of life in the Lardeau. Alas, theft and a fire in 1927 destroyed many of her exposures, otherwise we would have had more than the three cursory surviving pictures of the Lardeau tram installations. The remainder of the Gunterman's photographs became a core of the Vancouver Public Library's Historic Photograph Collection.[21]

Triune Mine

Another mine (and tramway) in the Lardeau was the Triune Mine. Located near the Silver Cup Mine, the Triune was above the tree line high on an exposed mountain top at 8000 feet. At the time it was considered "a very remote mine with the adit being atop a mountain, well above the tree line...." While the normal access via Ten Mile and the mountain paths to Silver Cup would serve adequately, the sheer height and location of the mine posed many problems.

Haulage was most difficult for bringing in wood and supplies; likewise the removal of ore was equally labourious. At first materials were packed in with horse and mule teams across the mountain scree and standing snow. It must be remembered that because of the altitude the mining season was very short—the mountain snow receded in late June, while the snow flew again in late September or October. The "Triune (was) considered (in 1902) the most inaccessible mine in the Lardeau."[22] The freight costs in 1900 for the Triune show both the value and the trouble involved: 20 tons were shipped while grossing $290 per ton before the $37 transport and smelting costs. To overcome the transport problems an aerial tram was ordered and "plans foran aerial tram were drawn up by Mr. Riblet."[23]

The parts were manufactured at Riblet's blacksmith and machine shop in Nelson. The local paper opined "This would be a difficult piece of work, even for Mr. Riblet."[24] The historian Milton Parent explains the tram construction at the Triune:

> All the tram machinery for the Triune had arrived by June, and ten men were immediately dispatched to clear the right-of way which would start from a point one and a half miles above the Ten Mile Wagon road. The system would be a vast improved. Nearly one and three quarters of a mile in length, the tram would be an even greater asset to the company, who had been paying dearly for the lengthy rawhide conveyancing of the their ore.[25]

> The tram took three months to complete and tallyed $20,000 in costs when finished in 1902. Even the weather at lower elevations caused difficulty. Due to the exceptional [winter] conditions in camp this spring, there will be some week's delay [in starting the mine this year.] Even the buckets for our tram are lying at Arrowhead and they will have to be shipped around to the Lardeau railway via Nelson.[26]

Presumably, there was ice on the North Arm of Upper Arrow Lake, though the exact nature of the conditions were not stated.

Yet all was not easy sailing with the tram for the next spring the weather was harsh. The tram stretched to elevations of 8000 feet using six to eight towers. However, other climatic difficulties came crashing down, this time in the form of a snowslide which obliterated a supply shed.

> Slides were everywhere, but most prevalent were on the Triune slope. An inspection team sent out in May found Number 3 and 7 towers on tram had been carried away, and in doing so the cable had jumped off and number of the remaining towers. And was, in many places, for hundreds of feet, buried in snow. Due to wind drifting, there was still seven feet of snow and the mine cabin had collapsed; the foreman said it would be June before it would be safe to bring in packhorses. Even then, any ore brought out would have to be by horses until the tram would was rebuilt and operational.[27]

Obviously, there were many problems operating a mine in such exposed conditions and at a great altitude. Even the pack animals were not immune and one was so badly hurt in slides that the packers had to dispatch it to the Elysian hay fields. The packers were nearly killed in the same slide too.

In 1905 the small, feeder tram was taken away by snow slides and the company did not have the money to replace it. This, combined with the earlier troubles, convinced the company of the perils of such high altitude mines. The Triune mine was abandoned.

With the demise of the Silver Cup and the Triune mines and the depleting ore bodies at the Nettie L, the boom in Ferguson had passed. Logging became the larger employer, and the hotels and mines were boarded up in Ferguson. Ferguson's debutante days were over.

Camborne

Camborne was another early mining camp near the head of the North Arm of Arrow Lake. It was promoted by Cory Methinick who was an early pioneer in the Lardeau region. He had earlier ran the Trout Lake Hotel and had owned a pioneering lake steamer. Branching out, he bought 188 acres of flat land on at the mouth of Pool Creek and the Fish River some six miles north of Ferguson.[27] While this spot was dark, sunless and forbidding, it had the benefit of being on the obverse side of the mountain where the Nettie L mine was. Above the Fish River confluence to the south loomed Lexington Mountain, where gold bearing quartz was located in 1899. Thus the Eva and the Oyster Criterion claims were staked and so it was hoped that Camborne would have a bright future.

Stores were established and lots were sold in the new town. All the while, prospectors continued to comb the mountains looking for mineral outcroppings. Assays from the Eva group yielded results of hundreds of dollars of ore content. As a result, the mining properties changed hands like poker chips. Meanwhile miners continued to hand develop the claims and sack and rawhide the ore down to the lakefront. By the first years of the new century newspaper and stock prospectuses waxed eloquently about the new mining region of Camborne. Two syndicates assembled the Eva and Oyster clusters of claims and thus they immediately began exploitation. Concentrators and stamp mills were installed at the foot of Lexington Mountain. Bunkhouses and drilling machinery were all installed to improve production. The Province agreed to fund an upgrade to the road. The canyon on Fish River necessitated some blasting and bridging.[28] The Oyster-Criterion Group brought in a ten stamp mill over the completed dirt road. The Camborne townsite was quite remote: it being some three and a half miles up the river from Beaton and the head of the North Arm of Arrow Lake.

Upper Terminal to the Oyster Riblet Tram at Camborne, BC. 1903. Photo: Vancouver Public Library No.86

With such a flurry of activity, aerial tramways were obvious choices to improve transport from the claims to the mill. A tram was ordered and for the Eva Group of mines for installation in the spring of 1902. It was 4200 feet long and of the "standard Riblet type...(with) a capacity of 100 tons daily and in excellent working order, being a difficult but well executed piece of work."[29] The adjacent Oyster-Criterion group, not to idly sit back, also contracted Riblet to construct a tram for them. It was

> 3500 feet in length leading to the ore bins and stamp mill. This tram, of the usual Riblet type, was just about completed and was first class in every way.[30]

Starting from the toe of Lexington Mountain—just west of the Eva tram—the Oyster Tram ascended the slash in the trees. The towers and stations were made from timber cut on site, while the ironmongery was brought in from shops in Nelson. Haulage was done by pack animals; both from the dock at Arrow Lake and from Camborne up the hill to the mine site. Gangs of men would cut trees and square them for assembly into towers and cribwork for terminal stations. Carpenters would manually drill bolt-holes and hand saw timbers. The animals would assist with ropes and tackle—to raise the towers and lift logs. Heavy bull wheels were rolled on the ground, like a wheel, with teams of horses into position.[31] Transporting the cables was always difficult for the teamsters. The loaded spools or cables were pushing the load limits for the horse drawn wagons. Also the bridges and state of the roads and inclines in the mining districts presented other difficulties. To transport the cables to the mine from Camborne, Daney, the horse packer unwound the cable from the spool and loaded two loops per side on each animal. In this way—with a line of animals—the entire length of cable could be packed in; while loading and unloading onto packtrains of horses was difficult and time consuming, it was a successful method of hauling the heavy, stiff cable in one piece.

With capital investment in the tens of thousands of dollars flowing into the mines, Camborne boomed. Houses, hotels, livery stables all grewup to assist the burgeoning population. A newspaper was established, too. One hotel in Camborne had the dubious pleasure of dispensing the most whiskey, for a hotel, in all of BC in 1906. Though it must be said for the all the vice, there was also virtue—the Lardeau

Horses packing tram cables to the Silver Dollar Mine at Camborne, BC. Photo: Vancouver Public Library, No. 20928

district organized dances, sports days, fraternal lodges, lectures, Victoria Day celebrations, and tobogganing. There was also the seasonal fishing, skiing, hiking and hunting recreational options too.

Yet with such meteoric growth, Camborne strangely had a similarly precipitous end. The year 1904 was the turning point in the town's fortunes. To start the dry room [a heated drying room to air wet mining clothes] burned to the ground by a faulty wood stove at the mine site.[32] Fortunately the fire was contained and prevented from spreading to other buildings. Another fire a few months later demolished the cookhouse.

Fire also visited the Eva mine that year, a forest fire burned out the top terminal of the Eva tram and mine buildings. Also the tram cables were scorched, ruining them by altering the temper in the steel. Thus, August brought a hive of activity with construction crews rebuilding the tram and upgrading the mine buildings and services.

Yet by the summer of 1905, both the Eva and the Oyster were silent. Mining production had ceased due to the vagaries of ore seams and business investment. It resumed again by the end of the year and was producing gold in the value of several thousand dollars.

Farther up Pool Creek were two other prospects: the Beatrice and the Silver Dollar Mine. The Silver Dollar Mine started production and even installed a 7000 foot aerial tram. This tram was built by the Crawford Tram Company of Nelson—an obscure company which seemed to produce few trams in relation to the activity of Riblet. After this citation, very little is heard of them. What did survive were two famous pictures of horses packing the tram cables to the Silver Dollar Mine.

Even as mining was having its ups and downs, logging was progressing in the area as the region was covered in prime, knotless cedar. Logging companies set up shop and a large mill was erected on the North Arm of Arrow Lakes at Comaplix.

In the summer of 1906, a summer rainstorm caused the rivers to wash out the Oyster's footbridge and water intake. After highgrading the seams, the mining syndicates became unwilling to forward development money to locate new seams in their mines, and the mines sat idle for months. By the end of 1906, only the Silver Cup Mine near Ferguson had any output. Inexorably, Camborne's inhabitants began to move away to look for other work.

The year 1907 was a disastrous year. It was a year of economic recession where investment money dried up after a ten-year boom period. As stated, various mines were playing out: whether in the Lardeau or in the Klondike. With the easy profits over, investors looked elsewhere—1907 is when the Guggenheims started work on the Copper River in Alaska, it is also the year when forestry first became a larger dollar value revenue source to the Province of BC than mining.

The catastrophic failure of the Silver Cup mill in Ferguson, combined with the foot dragging by the railways to expand lines in the Lardeau or connect the railways with their mainlines chilled investment like a winter blizzard. Thus Camborne began a quick decline.

Part of the problem was geologic. The ores of the Lardeau contained a high content of zinc. At the time the smelter in Trail was unable to use that grey metal and charged a fee for its removal. This is in addition to the extra freight one would pay on the concentrates. As a result the ores were uneconomic considering the large distances needed to travel and the frequent transhipments required for the journey. In 1916 Cominco developed the Electrolytic process to refine zinc at the smelter at Trail, and it became a valuable material and the Kootenays were a large producer for the war. But that was in the future and it was something that Camborne did not have.

Teddy Glacier

High on a mountain, some fourteen miles northest of Camborne lay the Teddy Glacier Mine. It was atop a mountain well above the tree line. First developed in 1925, the extreme remoteness of the site hampered progress. Caterpillar tractors, men and horse teams were used to sleigh in supplies. A Riblet aerial tram was even contemplated, though the exposed terrain and colossal expense of such an undertaking prevented its inception. In an attempt to make the operation workable, supplies were even dropped by aircraft. Despite a large ore body, the perils of transportation prevented the exploitation of this mine until the 1960's when miners better equipped with bulldozers punched a road in.

Meridian Mine

Camborne resorted to its somnolent, sylvan ways for the next twenty years. Only the Great Depression drove desperate men and women from the settled, civilized regions back to the hinterlands. The demand for gold opened up new mines in Zeballos, Bralorne, Bridge River and Germanson Landing. Old mining regions were revisited; driven by the price of gold.

The Lardeau Gold and Silver Mining Company was incorporated and, by 1932, it bought up the Beatrice mining claims on Pool Creek. It rebuilt the Oyster Mill, the old one having been flattened in the heavy snows, while the Meridian Mining Co. rebuilt the Eva and Oyster operations. The Oyster-Criterion aerial tram was reconstructed and put back into operation. The Riblet Tramway Company itself was in charge of installing the 3500 foot cableway. Most probably, the old trams were salvaged and pieces assembled into one working link. It ran for a few years until the Meridian Mine stopped operation in 1936.

From the Japanese Internment to Today

Camborne was one of the internment camps in the Second World War when people of Japanese ancestry—often Canadian born—were arrested and moved inland after the Imperial Japanese Navy's attack on Pearl Harbor. Tarpaper barracks were erected to house the people when they could not find room in the Edwardian Mine buildings. Camborne was just one internment camp of many: Sandon, Greenwood, and Slocan City were other locations as well.

Today, Camborne has only a few foundations and scrap iron leavings. Vandals have put flame to the remains of the concentrator resulting in only cement and iron traces amid a desolate landscape of tailings ponds. Furthering this devastation are the loggers who are in the process of burying their wood waste over the tailing ponds; around this scene are the clearcuts from where the bark, branches and stumps originate. An Edwardian office building survived until the mid-1990's but it seems to have succumbed to the elements or the hand of heavy equipment.

17. Southern Kootenay Lake Mines

Ainsworth and the Highland Mine

The panoramic pleasure spot of Ainsworth has a hot spring, which naturally drew visitors a century past—as it does to this day. In the process, prospectors searched the area and located silver outcroppings. Thus Ainsworth camp became a very early mining outpost on Kootenay Lake. "The Highland vein outcrops on top of a high bluff to the southwest of the village of Ainsworth. The claim was staked in 1893 by M. Stephenson."[33] A large mill was built at the lakeshore and Hallidie tramway was built to connect the mill with the mine in 1901. It is known that the Hallidie ran parallel to a Riblet design several years later but very little else comes to light about these cable systems. The Highland Mining and Milling Company operated the Hallidie system, while the adjoining Number 1 Mine used the Riblet link. The Riblet tram worked untl 1926.

Granite-Poorman

Eight miles west of Nelson, on Eagle Creek, was the Granite-Poorman Mine. It was discovered early in 1888 by I. Neil during the original Hall Mines rush and Nelson's establishment. "The mine is one of the oldest, and greatest producers in the District, producing intermittently from 1890 to 1954."[34] Its output was 4500 pounds of gold and 19,000 pounds of silver over this period. The Canadian Pacific Railway line from Nelson to Trail went along the south side of the lakeshore until it crossed the Kootenay River on a large bridge.
Transport to the mine was not a problem for a Riblet tram connected the mine to the railway siding about 1903.
While the 1918 *Report* notes on mine expansion—"The Eureka Copper Mines, Limited, is building another tramway, 3000 feet extended from the Granite-Poorman tramway to the railway track and on its completion expects to ship about 50 tons per day."[35]

Molly Gibson Mine

Roughly some twenty two miles north of Nelson was the rugged Molly Gibson mine. After a ten-mile trip up Kokanee Creek by road,

Twin Tram Towers: Riblet No 1 Mine Tram at right, Hallidie tram at work for the Highland Mine at left. Photo at Ainsworth, BC about 1912.

an aerial tramway took you to the 6900 foot level. The mine ascended higher inside the mountain. As the workings were above the tree line, it was a very exposed mine.

The mine was owned by the LaPlata Mining Company and was later sold to Cominco by the teens. The hundred ton capacity mill processed 20,000 tons of ore in 1907. The one and a half mile long tramway was put in at the same time. That was the first year of operation. This yielded a "gross value of $120,000" before a 61 man payroll and other costs.[36]

Angus Davis, the famed Cominco mining engineer, wrote in a history:

> In 1911 I was in charge of Cominco's Molly Gibson mine near Nelson. The mine lies at the head of Kokanee Creek, which, flowing from the north, enters the West Arm of Kootenay Lake about eleven miles east of Nelson. We had a ten mile road up the creek to the mill which was an elevation of about five thousand feet about sea level and with a seven thousand foot aerial tramway connecting the mill with the mine. In my time the elevation of the lowest tunnel up there, which was the lowest working was at an elevation of 8300 feet. Everything was above the timberline and there were slides all over the place.[37]

Indeed a snowslide slammed into the bunkhouse on Christmas Eve night 1902, and carried the men and structure over the hill well down into the treeline. Nine men were killed in the disaster.

The remoteness of the mountain mine site hampered development in other ways. Water had to be found because the tools were at first water powered which was a drawback when the water ran dry in the summer. Later, a Fairbanks Morse semi-diesel engine was installed to replace the uneven waterpower. "Furthermore, all our mine timber had to be taken up to the mine on a timber carrier on the aerial tramway, a slow and tedious process."[38]

Another difficulty was securing a reliable cook, for without decent food, Davis he quickly soured on the isolation. One "gutrobber" went mad and so Davis had to place another.

> I sent an old German up there to cook. It was winter and he had to ride the tram. Now, at the upper terminal there were about twelve cross timbers with sheaves on them to take care of the running rope. These just nicely cleared the buckets as they entered the terminal. This old fellow, as he entered the terminal, got excited and stuck his head up a little too far and got hit by one of the beams and then, up and down went his head, for he wouldn't keep it permanently down, so each of these timbers gave him a bang as he went by. I had him there all winter as he was scared to death of the snow slides and he could not snow shoe anyway. As for the tram, one could not get him within gunshot of it. He was a good cook too....[9]

Davis gives us very detailed descriptions of the perils of tramways, and while he does not say the maker of the tram, he implies that it is a Riblet product.

> Still on the subject of the tram, riding up on it one day I happened to be in the bucket immediately following the timber carrier, which had a big stull [timber] lashed to it, and I noticed that each time the carrier passed over the tower the stull would hitch a little further sideways, for it was improperly tied on, and finally, there it was, absolutely at right angles to the line of travel. There I sat, watching it approach the next tower and half scared to death, for a wreck somehow now seemed inevitable. The running rope however took the strain and although the bucket I was in oscillated violently up an down for a few times, nothing parted company and I was safe.[40]

Everyday life in the Kootenay Mines was not without its adventures. Davis expands upon hiking trips over the mountains to Ainsworth, and the unique Taylor air compressor on Coffee Creek—a 200 foot raise in the rock and used creek water and a height gradient to pressurize air. Davis would later continue his adventures with the mining boom in Stewart.

Davis wrote one of the finest tributes to the Riblet Brothers.

> **As more ore was opened up in the Slocan, aerial tramways were developed to relieve the pressure. Around 1898, Royal N. Riblet of Spokane and his brother [Byron], both young engineers, [sic] turned up in the Slocan having developed a new type of aerial tramway, "the Riblet tramway" they called it, and it was particularly adapted to the rugged country and the high hills of the Slocan. The Kootenay owes a debt to the Riblets as their aerial tramway was away better in cheapness and efficiency than anything so far developed throughout the world along those lines. Seizing the opportunity, they built thirty of forty in short order. They were anywhere from a mile or two miles or more in length, and they unlocked all kinds of ore that would have otherwise stayed in the hills for further prolonged periods.**[41]

The LaPlata Mining Company went bankrupt in 1908 and the property unworked until Cominco bought it in 1910. They later installed a second, five mile tramway to eliminate bad parts of the road.[42]

Notes—4

1. Theodore Otto worked with Adolf Bleichert perfecting bicable designs in Germany in 1874. Bleichert later became the world leader in mining trams, and it is still in business today making material handling systems for factories. Otto was a lesser competitor in the North American tramway business and formed partnerships with Fraser and Chalmers the Wisconsin smelting retort makers (later Allis Chalmers.) The Broderick and Bascom Tramway Company was also allied with Otto and became an agent for their wares in North America. For information on Otto Tramways (see Trennent. *Riding the High Wire: Aerial Mine Tramways of the West.)* For history on the Lanark Mine (see BC Minister of Mines website "MINFILE.")

2. *Report of the Minister of Mines.* 1907. p. K117.

3. Milton Parent. *Circle of Silver.* Arrow Lakes Historical Society. Nakusp. 2001. p. 9.

4. Ibid. p. 11.

5. Ibid. p. 47.

6. Ibid. p. 34.

7. Pettipiece was rabidly socialist and railed against the world, capitalists and Orientals; while he was for the worker, eight hour day, and the town of Ferguson. In his writings he could be clever, he could be tiresome and he could be descriptive. He was in mortal struggle against the other paper, *The Trout Lake Topic* with some wonderful battles in print. Pettipiece later moved to Vancouver worked for the *Daily Province* and was elected to city council. Ibid. p. 59.

8. It must be remembered that the world was going through convulsions about another mining district at this time-that of the Klondike Gold Rush. With the discovery of gold on Bonanza Creek, Dawson City and the arrival of the steamer *Portland* in Seattle with holds of gold, a pandemonium ensued. What with the narrow focus of business and media, the Lardeau was placed further down the dance card as the world became familiar with the Chilkoot, sourdough bread, Soapy Smith, Jack London, Robert Service and the inimitable Pierre Berton. Ibid. p. 59.

9. *Railway and Shipping World.* May 1899. As quoted in Wilkie and Turner. The *Skyline Limited.* Sono Nis, Victoria. 1993. p. 175.

10 Ibid. p. 175.

11 As transportation is symbiotic-the railways, steamboats andaerial tramways each generated freight for the other. In this way steamboats and steamboat captains of the Kootenay became famous. The tales, anecdotes and names of boats fill the pages of several books. Yet Kootenay steamboats have an Imperial reach: some CPR steamboat captains saw service with the Royal Navy Volunteer Reserve in the First World War. Through this people, from BC, saw service in the Middle East. During the First War the British were trying to dislodge the Ottomans from Mesopotamia. At the time Turkey was seen as a threat to India. In 1915, General Townshend was dispatched from India with Indian and British Troops to take Baghdad. They marched on foot several hundred miles and reached the town of Kut. There they were besieged by a large army of Ottoman Turks. As a result they dug in and were harassed. The Royal Navy assisted the army with armed riverboats, some where skippered by the CPR captains. These ships were the famed Insect Class gunboats. The campaign failed as the troops were surrounded and beleaguered for months. Between the Turkish shells and disease, the force wasted away until they eventually surrendered. The Allied prisoners of war were then forced to labour on the then unfinished Berlin-to-Baghdad railway.

12 George Attwood. "Plant for the Handling and Treatment of Ores Silver Cup and Nettie L mines, BC" *Proceedings of the Institute of Civil Engineers*. London. 1903.

13 Ibid. p. 297.

14 Ibid. p. 299.

15 Ibid. p. 299.

16 Parent. *Circle of Silver* p. 127.

17 Ibid. p. 137.

18 Ibid. p. 118.

19 Ibid. p. 117.

20 Ibid. p. 139.

21 For many years the VPL had vast enlargements of the Camborne photographs at the Robson street location of its archive. They dominated the room. (See Robineau. *Flapjacks and Photographs A History of Mattie Gunterman: camp cook and photographer.*)

22 Parent. *Circle of Silver.* p. 126.

23 Ibid. p. 156.

24 Ibid. p. 156.

25 Ibid. p. 117.

26 Ibid. p. 126. As quoted by the Triune mine manager to a reporter of the day.

27 To this day avalanches and slides in the Kootenays still cause death and damage-the toll includes sons of ex-Prime Ministers, school groups and ski fans such as Geoff Liedel. Ibid. p. 127.

28 Ibid. p. 62.

29 The Fish River is a wild one. Spring runoff brings huge torrents, while the Canyon itself is not for the faint of heart. All machinery, food, supplied and liquor were brought in via this road. Ibid. p. 96.

30 The *Report of the Minister of Mines* lists annually all development work on all mines in the Province. Sadly it does not always list the installation of tramways, the individual maker, or the quality of work. *Report of the Minister of Mines* 1903. p. 128.

31 *Report of the Minister of Mines.* 1903. p. 130.

32 Milton Parent. Op. Cit. p. 148.

33 For details see www.em.gov.bc.ca/cf/minfile. The online summarized *Report to the Minister to Mines* database for the Highlander Mine.

34 Source of the Granite Poorman data-www.em.gov.bc.ca/cf/minfile for Kenville.

35 *Report of the Minister of Mines.* 1918. p. F194.

36 "Granite Poorman." *Report of the Minister of Mines.* 1908. p. L104.

37 Clara Graham and Angus Davis. *Kootenay Yesterdays*, Vol.4. Alexander Nichols Press. Vancouver. 1976. p. 163.

38 Ibid. p. 163.

39 Ibid. p. 164.

[40] Ibid. p. 164.

[41] Even the educated Davis is bamboozled by Royal Riblet's showmanship and attributed most of the credit to him. All the same, Davis does acknowledge the importance of Byron Riblet's products. (See Clara Graham and Angus Davis. Op. Cit. p. 158.)

[42] MINFILE Reports for "Molly Gibson." www.em.gov.bc.ca

South West Coast

18. Mount Sicker and the Tyee Mine

North of Duncan on Vancouver Island is the prominent Mount Prevost—an isosceles shaped interruption. Adjacent to it is Mount Sicker—and under Mount Sicker were important copper deposits. With the coming of the European, the Island became set for exploitation. What was needed was easy communication. To this end, the CPR and Dunsmuir Family commenced building the Esquimalt and Nanaimo Railway to fulfill its contract for a transcontinental from Montreal to Victoria. Thus, the line was built through Esquimalt, the Western Communities, across the picturesque Malahat, Shawnigan Lake, Cowichan Valley, Chemanius, Ladysmith, and Nanaimo. The line was later extended north to Qualicum Beach, Fanny Bay, and Courtenay. The grade was finished to Campbell River, but track was never laid. With one hundred miles of line came a sizable land grant. Almost one quarter of the Island's land area: a block of land from Nanaimo to Victoria and Inland to the mountains. With the land came the timber, mineral and water rights. It was an incredible deal. The CPR subsequently sold off some of the land; the Dunsmuir's used some for their coal operations; loggers bought other tracts for the valuable old-growth fir; and settlers bought some for homesteading.

With the railway wholesale exploitation of the land commenced. The mineral deposits were thus extracted, as at Cassidy and Extension. Also within the land grant, only several miles from the Esquimalt and Nanaimo Railway, lay the copper deposits on Mount Sicker. The Lenora claims were established in 1897 by Harry Smith. Subsequently, underground workings set up. Problems in development ensued and the claims were sold to Henry Croft, a relation of James Dunsmuir. Hand mining continued, at the same time a haul road, town and school were established by 1900. To access the mines, a narrow gauge railway was built and two brand new Shay engines puffed their way up and down the hill.

Bunkhouses and a hotel were built and by 1900 the mines on Mount Sicker were a going concern. Another mine nearby was the Tyee. It was at the southern head of Copper Canyon, overlooking the Lenora operations. Smith had cashed out his claim to the Mt. Sicker and BC Development Company, who were, in turn, reorganized into the Tyee Copper Company using English money. Unlike the rough and ready handmining at the Lenora, the Tyee brought in large, new machinery to work their mine. This included two large, horizontal return flue boilers and concomitant brickwork, a steam-air compressor for the rock drills, and large steam hoist to draw the ore kibbles.

The Tyee mine had been shipping ore using horse wagons to the E and N railhead at the foot of the mountain. Then the trains hauled it to Ladysmith, where it was stowed on coasters for the trip to the smelter at Tacoma. Yields were very high: the ore content was 8% copper.

Rather than contract with its neighbour and use their railway, Tyee's owners decided to build an aerial tram. To improve their transport situation the Tyee Copper Company hired Riblet to build an aerial tramway from the 1700 foot level of Mt. Sicker over a ridge of the 2200 foot mountain down to the foot of the mountain at near sea level. Its ferrous filaments were one and three quarter inch in diameter on the heavy side, and seven eighths inch on the light side and stretched over three miles long.[1] An ore bin was built at the E and N Railway siding to store the material while collecting carlots. Capacity was rated at 5000 tons per month.

In April, 1902, the *Nelson Daily News* remarked:

R. N Riblet returned on Friday evening from Vancouver Island, where he had been overseeing work on the tramway from the Tyee Mine. Mr. Riblet states that one mile of the right of way line had been cut. The total length would be three miles. Most of the machinery of the tram will be manufactured at Nelson, the shops of the company now being very busy turning out the necessary parts. The season at the coast is very backward, the weather remaining cool with more or less rain every day.[2]

The tramway was completed in the fall of 1902.

With the tram set up and the machinery installed, the Tyee hit full production. The next year it shipped 42,000 tons of ore returning over a half million dollars in profit their income was yielded from hauling ore via the tramway from the neighbouring Richard III mine.

In 1905 the Tyee mine was able to maintain shipments of 2500 to 3000 tons per month and although their main shaft was down 1000 feet there was still nothing in sight but low grade ore. During the year the Tyee shipped 31,900 tons of ore to their smelter and recovered 2,688,945 pounds of copper, 87,028 ounces of silver and 5003 ounces of gold for a total of $525,000.[3]

To take advantage of the numerous deposits of coal, a mineral smelter was built at Crofton Bay. The Northwestern Smelting and Refining Company financed and built the large facility. "The Smelter contained three 200 HP boilers, a 500 HP Corliss engine, (and) a 450 ton water jacketed furnace."[4] A massive twelve foot diameter chimney towered 150 feet over the bay, spewing its noxious, sulphurous gas. The smelter was built to treat the ores from Mount Sicker. However, it also offered its services on the open market and smelted other people's ores as well. Not long after it was built, the company ran into financial difficulty and sold the smelter to the Britannia Milling and Smelting Company.

As usual the Tyee Copper Company and the Lenora and Mt. Sicker company did not share resources. They ended up building two separate smelters. "The English owners decided to build their own smelter." In 1901 James Dunsmuir was brought in as a partner in the Tyee venture, and he "provided a site on Oyster Bay directly in front of the new town of Ladysmith."[4] With half a mile of waterfront to provide deep-sea access, and neatly adjoining the coal mines of Nanaimo and Cassidy, Ladysmith was an ideal spot. Coal and ore were conveyed eight miles from Mt. Sicker by the Esquimalt and Nanaimo Railway.

The smelting retort was in operation by the end of 1902. Cordword was piled on the ore which was heap roasted in fires. Then the ore went to the 150 ton refractory smelter which removed the trace gold and silver and reduced the copper to blister copper. Still impure, the semi-refined ore went to Tacoma for finishing. A 80 HP boiler and Corliss engine provided the power for the furnace blower.

The smelter only operated for nine years when the ore at Mount Sicker gave out. Ladysmith then became a principal lumbering town with sawmilling at Chemanius and Oyster Bay being a railway and logboom staging area.

> A government inspector made note during a tour of the group [of Mt Sicker mines] that it would have been much to their mutual advantage if all three mines could have been worked as one with need for only one haulage system and smelter.[6]

In the end, the financial overextension, and drainage problems in the mines hastened their demise. Thus, by 1907, production dropped off at Mount Sicker and the assets of town, railway and mining equipment were subsequently sold. The Crofton smelter eventually was

Photo of the Tyee Mine tramway at Mt. Sicker, Vancouver Island. Photo Engineering and Mining World Magazine. 1904.

demolished and the site became the present pulp mill in the 1950's.

In 1918, the BC Molybdenum Company opened a mine near Cowichan Lake. It had an aerial tramway. It is possible that this used parts salvaged from the Tyee Mine, though no corroborating details have been found to support this.

19. Britannia Beach, B.C.

The mining community at Britannia Beach, B.C. is an outlier. Contrary to the trends in BC mining, it had a long lifespan. Unlike most of these old mining communities, it did not disappear into the dusty pages of history. It blossomed into the BC Museum of Mining. Similarily, the mining site is not a ridiculously remote one situated hundreds of miles from supply points and then perched precipitously high on treed mountainsides; rather, it is a mere thirty miles from the metropolis of Vancouver. It is serviced by Highway 99—the congested, dangerous and winding path to the ski mecca of Whistler.

A Doctor Forbes worked in the area in the 1880's attempting to reduce the suffering of the unfortunate smallpox ravaged aboriginals. Copper ore was brought to his attention by the Salish natives, thus he prospected for that metal north of Furry Creek.[1] Late in the day on one tiring, fruitless trip scouting in the steep mountains, the doctor shot a buck for provisions. In the process of dressing the meat, the doctor noticed that the buck had died on a mineral vein. (Animals seem to feature in mineral prospecting; be it in Wardner, Idaho, Nelson, BC or Tonopah, Nevada.)

The claim was located by Doctor A. Forbes in 1888. The Britannia vein was some three miles from the beach at Howe Sound. It was not until 1901 that industrialization and production of the property began. "George Robinson, a mining engineer from Butte, Montana, visited the property. Upon inspection, he concluded that it was well suited for a mining operation."[2] With New York money, the Howe Sound Company was formed. It became the parent company and was principle operator at Britannia and Mt Sicker and other mining ventures. The actual mining was undertaken and organized by the business unit Britannia Mining and Smelting Company.

Up on the hill—at the Jane Basin, the adit was developed as well as bunkhouses,compressor stations, machine and blacksmith shops, cook and bunkhouses for the miners and the company. While at the beach site, the town was developed with housing, company store, wharf and a concentrator. To connect the two locations Riblet was contracted in build an aerial tramway. The total length was 16,500 feet, built in two sections with an angle station. The Britannia tram used $1^3/8$ inch diameter track rope in its bicable setup and was capable of moving 50 tons of ore per hour. It used the standard Riblet thousand pound ore bucket. There were between 50 and 100 buckets moving on the line depending upon demand.[3]

The ore was hauled by surface tram from the mine to a storage ore bin at the Jane Basin. It was then loaded automatically onto the buckets by the Riblet patent loader. These brought the ore through the slash in the Coastal BC forest, and down the steep bank to the beach. There it was fed to another ore bin above the Number 1 concentrator. Finally it was fed into a crusher for processing in the mill. The standard Riblet towers were constructed of locally cut timber and erected as needed. Final cost of the tramway came to $75,000 when it was put into operation in 1903.[4]

A Riblet Tramway at work near Britannia.

A diarist of the time noted a death by the Britannia tram. A new doctor was travelling to the mine by the tramway, and despite the tramcrew's repeated warnings to watch his head on the towers, somehow he lifted his cranium when the bucket was passing through a tower. He eventually died from the trauma inflicted by the tramway tower beam.[5] One of the few references to trams in the 1915 *Report of the Minister of Mines* noted that the electrician's helper J. Peddel was "caught in a tramway bucket at station and badly mangled; instant death."[8] Mining was a dangerous business for all involved. The back pages of the Report, where the fatalities are listed, make for some grim reading; the coal mines always had the highest death toll.

Perils of a different nature, often visited the nascent mining camp of Britannia. One can argue it is one of the most cursed mining camps in the province. In 1906 a snowslide tumbled down and demolished a cabin at the Jane Basin camp. Six men were killed as a result. A larger mudslide in 1916 killed 57 people. And this was not the end of the Grim Reaper's dealings at Britannia.

In 1921, a period of heavy rains caused extensive, tragic flooding. It seems that a mining railway crossed a creek. In the process, the railway embankment curtailed the natural passage of the waters, as the railway built a timber crib culvert which the railway could pass over the creek. Debris in the creek, such as large logs and stumps, would be washed down by the flooding and impeded by the small watercourse at the timber crib. During this November flood men were stationed at the crib to clear debris from the bridge. Yet these men were sent home at dusk. In the night, logs blocked the crib culvert and water backed up in a natural dam. As the lake increased in size, the water pressure increased on the railway embankment, eventually it became too much and the dam gave way. The torrent of water tore down the hill and smashed into the town. Houses were demolished and washed into Howe Sound. The death toll was 37 people and many more injured.

The effects of the First World War and the Spanish Flu epidemic of 1918 took their toll, although these were not unique to Britannia Beach. Bruce Ramsay explained the effects of the Influenza:

> **Men died like flies in the bunkhouses, and the dance hall at the townsite became a temporary hospital. At the beach, Dr. Roberts, the camp medical officer, had his hands full. The hospital was full, every house had someone sick in it, and his task was madeeven more difficult when the nurse at the Townsite succumbed to the ravages of the 'flu'. The best the doctor could do was to give the men plenty of whiskey and hope for the best. At the Townsite, Bob Ryan, a first aid man, took over the job of nursing the camp back to health, and at the beach Yip Bing, the Chinese boy who worked in the store, made and delivered on his little wagon pots and pots of soup, earning him the title fo "Doctor YB."**
>
> **At night the aerial tramway became a funeral cortege as bodies, sometimes as many as a dozen at a time, were brought down to await the steamer for Vancouver. It is not known how many died in the Flu epidemic at Britannia, but estimates range from 40 to well over the hundred mark.**[7]

Despite these horrific setbacks, the Britannia mining community settled into a long period of quiet productivity. With a capable manager at the helm copper output increased. It was the largest copper producer in the British Empire. Only the copper mines of Arizona and Utah were larger.

The Britannia aerial tram was a success and operated for about a dozen years. Afterwards, increased mining production necessitated the transport of ore in more voluminous quantities. A surface funicular was built during the First World War. As the company had reaped most of the high grade ore, development started with the lower grade material. To capitalize on the situation, a newer floatation mill was built on the eastern side of the first mill. "Production increased tenfold, from 200 tons per day to 2000."[8] Yet, this mill burned to the ground in 1921. A larger, fireproof cement and steel replacement—Number 3—was subsequently constructed. This is the large building that stands to this day.

To feed the mill, the mine tunnels were eventually linked up inside the mine. Ore was ferried to re-load stations by skips and then the ore was hauled directly to the mill with an electric railway. The labourious and involved surface tramway—both aerial and inclined—were now obsolete. Britannia was very much a company town.[9] While the bright lights of Vancouver beckoned and glowed a mere 30 miles away, transport was difficult. Both the Pacific Great Eastern railway and the highway on the eastern shore of Howe Sound were not completed until the 1950's. Prewar communication was via the scheduled Union Steamship sailings. As a result, the mining company had a larger measure of control than most modern people live with. The miners and workers lived in company owned housing. Social mores trickled down from the management, and all were expected to follow. Purchases were made at the sole and company owned store. One anecdote from Britannia was the account of the miner asking for his time so he could quit as he was fed up with the Company, to which the Superintendent retorted—in a manner reminiscent of "Tennessee" Ernie Ford—"He can't quit, he owes too much to the store."

The Second World War continued the demand for metals. A strike in 1946 signalled some internal company troubles; while the recession of 1957 caused the liquidation of the Britannia Mining and Smelting Company. The Jane Adit or Mount Sheer townsite was abandoned. Later the company assets were later purchased by the Anaconda Copper Co. of Montana and the mine resumed production. More houses were built near the beach and the future seemed rosy.

However, by 1974 competition from open pit mines and a new socialist Provincial government in Victoria modernised the Edwardian mining laws, which forced the conditions for the final closure of copper mining at Britannia Beach. A development company bought the land, and the Britannia Beach Historical society effected the opening of the BC Museum of Mining using the buildings and old operations. Underground mining tours are run for the benefit of the public.

Other pressures have impacted on Britannia Beach—the urban sprawl from Vancouver has made Howe Sound into desirable real estate and a golfing destination, while the famed ski resort of Whistler has further increased population and development pressure on the region.

Furthermore, the legacy of seventy years of mining has left a large environmental footprint. The broken leftover rock from mining combined with rainwater, produces acid runoff. In keeping with this legacy, Britannia is one of the most polluted locations in BC. The Provincial Ministry of Environment is seeking solutions to ameliorate the problems of acid leachate in Britannia Creek and Howe Sound. In 2001, Victoria started a $25 million agreement with the property owners and Anaconda Copper (the last mining operator) to attempt to introduce lime into the water to reduce the acidity. Consequently, Britannia Beach's history is still with us today.

Notes—5

[1] White and Wilkie. *Shays on the Switchbacks*. BC Railway Historical Association. Victoria, 1963.

[2] Ibid. p. 22.

[3] *Nelson Daily News*. April 1902.

[4] White and Wilkie. Op. Cit. p. 39.

[5] Ibid. p. 17.

[6] For a brief history of smelting and the formation of Ladysmith, BC. (See Verchere.com/smelting.htm)

[7] White. Op. Cit. p.33.

[8] Mullan, M. Britannia-*The Story of a BC Mine*. www.em.gov.bc.ca/Mining/Geolsurv/Publications/Openfiles/OF 1992-16-Pioneer/Britannia.

[9] Ramsay, Bruce. *Mining in Focus*. Agency Press, Vancouver, BC 1968.

[10] This was a literate journal by the wife of a traveling Colorado Miner. See Harriet Fish Backus. *Tomboy Bride*. As quoted in Trennert, *Riding the High Wire*. p. 63.

[11] *Report of the Minister of Mines*. 1915. p. 423.

[12] Bruce Ramsay. *Britannia–The Story of a Mine*. p. 55.

[13] Mullan. Op. Cit. p. 7.

[14] There is an amusing story from Phyllis Munday, the grand dame of BC Mountaineering, in the 1920's she hiked around the Howe Sound Peaks using Mount Sheer mining camp as a base. She was travelling with her infant child and basic needs came up. Lacking a wash-basin at one of the cabins, she used a large mining breadpan to bathe the child. To keep the child warm she perched the pan on the ovendoor of the wood stove. A miner came in, saw the scene and quickly fled fearing an imagined baste-and-baking of the baby.

The Boundary Region

20. Grand Forks, B.C.

As part of the development of the Granby Mining, Smelting and Power Company, industrial facilities were erected at Grand Forks, BC. The copper ore of the district was rich, with mines at Greenwood and Phoenix, and smelting was soon in order. The Granby Company elected to raise a smelter at Grand Forks, and began construction in 1899. With $300,000 in capital, the company was a large undertaking. It had some 300 men employed in the smelter and adjoining hydro dam construction. Byron Riblet was contracted to supervise the building of the Granby Smelter dam across the North Fork of the Kettle River at Grand Forks. The dam was needed to provide power and source of water for cooling the retorts of the Bessemer Converters. A reporter of the day captured the scene:

> From the dam across the Kettle River to the superintendent's office, a distance of nearly one mile, there are gangs of workmen, puffing engines, piles of lumber, loading teams and everything that makes an inspiring spectacle of thrift and industrial activity.[1]

The dam itself was a timber crib structure, so common in the pioneer era. Dams of the day were not the latter grand cement or earth edifices that national investment allowed, rather, they were more modest timber weirs. At Granby's dam water was piped in a large flume to the powerhouse. "The flume will be the largest and longest [then] constructed in British Columbia."[2]

The smelter itself contained many structures of brick including engine house, dust chambers, assay building, chimney and residences. Other large wooden buildings including the crushing plant, and smelting building were built as well. The smelter was 'blown in' in 1900. And like the smelters at Nelson, Northport, Boundary Falls and Greenwood, they operated for a few years before being superseded by the large smelting complex at Trail.

21. Phoenix, B.C.

For the traveler the Boundary Region of BC is beautiful country. Yet the Boundary Region was valuable too. Successful copper mines were in operation just south at Republic in Washington State. This hinted at tremendous wealth to the north. Early on, in 1891, the No. 7 Mine was staked near the future city of Greenwood. In time, the Boundary Mines Company was set up with New York money.

Production ensued with hand methods and draft animals. By 1899, transport improved with the arrival of the Canadian Pacific Railway's Columbia and Western branch line.

Soon, by 1901, the BC Copper Company had a smelting operating in Greenwood and it was treating the No. 7 ores amongst other mines rock. Output lapsed on the No. 7, until the Consolidated Mining and Smelting Company took an interest. They built a two-mile aerial tram from the mine to Boundary Falls, a few miles west of Greenwood.

At Phoenix, high above Greenwood City, a large copper deposit had been located and became set for exploitation by 1898. The MacKenzie and Mann interests developed the property and soon were shipping ore to the Hall Smelter in Nelson, and the Granby Smelter at Grand Forks. Phoenix itself sat over the large copper deposit. Twenty-six separate companies staked the ground and drove shafts below to extract the ore. The claims were registered with florid names such as Knob Hill, Old Ironsides, Victoria, Phoenix, Brooklyn, Stemwinder, Snowshoe, Oro Denoro, Fourth of July and Aetna. Above, some ten thousand beings, toiled, built, and lived. Periodically, the ground would subside from the subterranean burrowing, and swallow a building. A fact of no matter to the pioneers, they would simply rebuild. It was a full ahead town. The ore was shipped by rail to the smelters at Greenwood, Trail, or Grand Forks.

Yet the industrial activity increased to a more harried pace. With the help of Spokane and Eastern Canadian financiers, the Granby Consolidated Mining, Smelting and Power Company was incorporated and set to work with a capital of $15 million. It became the preeminent company in the region. Soon, the company's Phoenix mines were yielding profits in the order of a million dollars to its investors. In turn, the Granby company became a large mining operator in Canada. The company underwrote the construction of the town. Phoenix contained all denominations of churches, schools, a hospital, drinking establishments and even a champion NHL hockey team who played on artificial ice.

With such a scale of money on the table, technical improvements were the norm. This included aerial tramways. An aerial tram was built by Riblet to haul material five miles from the Lone Star Mine around 1909. That mine was within Washington State, though the ore was hauled down the mountain to the facilities at Boundary Falls, just west of Greenwood. Unfortunately, again information on this installation is sparse. While much of the details are unclear about the two aerial trams in the Greenwood vicinity—the Consolidated Mining tram and the Lone Star—it is known that Leschen built one of them.

Leschen's effort was configured in a to and fro skip type, carrier tram to haul blasting powder from Greenwood to Phoenix in the first decade of the twentieth century. It was a different design than the mining carriers as it had two track cables to support the large box. Obviously, the engineers wanted to separate the transport systems in case of disaster with the explosive. Alas, it is not known which of the two trams Leschen built.

A photo of the dynamite aerial tram from Greenwood to Phoenix, BC about 1907. It was one of three trams in the BC Boundary region in the First World War era.

Phoenix was one of the larger mining camps in the region and worked until the end of the First World War. It had large Lorain Steam shovels loading specialized ore cars, which delivered 7000 tons of ore per day to the two railways.[3] There were early attempts at open pit mining, too. However, with the drop in ore prices at the end of the war, and the competition from the Guggenheim operations in Alaska, Utah and Chile, mining at Phoenix ceased. The operators ventured elsewhere—the Granby Company to Anyox; the BC Copper Company to Allenby at Princeton.

In a short period of time, the town of Phoenix completely disappeared as its buildings and materials were salvaged. So too was the fate of the smelting equipment at Boundary Falls, Greenwood and their espaliered growths of aerial tramway. The Phoenix townsite was later turned into an open pit mine in the. Today, only the war memorial and a few foundations stand.

At the end of the First World War the neighbouring Rock Candy mine entered production. It was another mine located to the north of

Greenwood and high on the mountain. To bring the ore down to the smelter, Riblet built a tramway to connect the two. The lifespan of the Rock Candy mine was not long, though a few remains of this Riblet link survive in out of the way places. Greenwood today, on the other hand, has re-entered the world's radar scopes as it was recently cast as a San Juan seaport in the cinemagraphical rendering of *Snow Falling on Cedars*. A fitting connection, when all things are considered, as the Boundary region was one of the Japanese Internment centres in the Second World War.

22. Salmo, B.C.

Salmo sits strategically in the Salmon Valley. A few dozen miles north up the Salmon Valley lies Ymir and Nelson; if one travels west via Fruitvale the path leads to Trail; east over the high pass is Creston, and if one heads south along the picturesque valley leads to the US. Yet Salmo is important for more than a crossroad Seven miles southeast of Salmo on Sheep Creek was the important Sheep Creek Mine.

The mining activity of Rossland and Nelson brought prospectors to the area. Of course the region on Kootenay Lake became a hive of activity, yet prospectors scoured the valleys of the region. The Sheep Creek mines were discovered quite late—1907—by two men (Horton and Benson). It yielded high-grade lead-zinc ore though the complexity of the ores reduced output due to the higher smelting costs.[4]

The mine site was reached by an easy road from Salmo. With a camp and concentrator just off the Sheep Creek valley floor, the ore deposit was located some 3000 feet up on the north side of the valley. Of course to access the ore and bring it out an aerial tram was constructed in 1911. Transport was "by means of an A. Leschen and Sons aerial tram 3700 feet in length with a vertical distance between loading and discharge being 1900 feet."[5] It should be remembered that Byron Riblet sold some tramway patents and became a chief engineer for Leschen from about 1903 to 1911. Thus this tram most likely was joint effort by both Leschen and Riblet.

The 1915 *Report of the Minister of Mines* goes on and states "the tram has a capacity of 100 tons and is operated by one man only, the bucket being automatically loaded and self dumping."[6] A 1912 photograph in a 1930 Leschen catalog shows a load of lumber bound for the mine on a carrier amid a denuded and greenless forest.

Nelson Machinery Company—the large used machinery dealer in Vancouver—acquired the Sheep Creek Tram and was attempting to sell it whole for reuse in 1956. They noted some more technical details of this Riblet/Leschen tram: track cable size one inch, running cable size 1/2 inch, bull wheel diameters—52 inch, trapezoidal bucket—23 inches deep and a four inch brake block on the running cable sheave. The tram was not re-sold and it subsequently went for scrap.

Judging from the dimensions, this was a medium sized tram, with a smaller bull wheel, span and cable size. Though this was not the smallest of trams, either a 'to and fro' single carrier or double jig back style. Nonetheless, it was a pity it was not kept.

The Sheep Creek mine had sporadic output owing to the geology. Cominco bought the property in 1927 and investigated its ore. After the Second World War production was really ramped up and a 1000 ton capacity mill was installed, the mine resumed production. Unlike so many other mines in this story, the mill is standing today and the mine is idle. However, considering the start, stop method of production from Sheep Creek mine, it is not unlikely that production could resume given the right market conditions.

A Leschen lever grip used to attach buckets to traction cables. Riblet used a spring system

23. Blakeburn, B.C.

In 1920, a new mine started up in BC. That operation was Coalmont, near Princeton. Princeton and Hedley were already thriving mining communities in their own right: the former with Copper Mountain and Allenby, and the latter with the Nickel Plate Mine at Hedley. Consequently, prospectors searched the region for other resources. Coal was discovered to the south of Tulameen River some fourteen miles west of Princeton, on Lodestone Mountain in 1905. A ground squirrel scratching near his burrow, caught the prospector's eye and drew attention to the presence of coal. This would eventually become the mines of Blakeburn.

Consequently, capitalists had optioned the Lodestone Mountain properties and had done some exploratory work. However, the distance and rustic transport methods were a hindrance and duly noted. "When transportation facilities reach this rich mineral country this coal will be of great commercial value."[7] The Columbia Coal and Coke company was subsequently formed and tried to exploit the coal from atop Lodestone Mountain in a location called Fraser's Gulch, in the end the coal turned out to be of poorer quality.[8] Operations were then moved down the mountain to 3700 feet, and nearer the canyon to a site with better grade coal.

In anticipation of the boom the Coalmont townsite was established in 1910 in the valley on the Tulameen River—at the foot of the mountain near where Granite Creek joined the Tulameen. Stores, schools and houses were erected, and building lots sold. At this time seventy men were at work in the mines, driving tunnels, establishing shops, and hauling materials via a tote road. In the process 50 tons had been shipped by 1913 while 4850 tons shipped the next year.[9]

At first the Columbia Coal and Coke planned a narrow gauge railway from the minesite on the North Fork of the Granite River down to what is now "Anchor Hill" overlooking Coalmont. The coal was to be carried down the hill on a funicular. The narrow gauge road was surveyed though the plan not effected. Alas, the financial malaise of 1913 and the transport problems stressed the aggrieved Columbia Coal and Coke Company and caused production to cease during the First World War.

In 1918 fresh money was brought in and the company reconstituted as the Coalmont Collieries Company with a capital of three million dollars. The Alberta butcher Pat Burns was a principal partner, and along with W. Blake Wilson serving as president, the company had a bright future.[10] On the mountain, the new mine site was re-christened

Blakeburn from a neologism of the two names.

Production was resumed and the tons of coal hauled out with primitive Federal Trucks over the winding narrow, canyon road. Transport would have to be improved have a successful mine. Thus an aerial tram was built.[11] The right of way was cleared of trees in 1919 and construction began.

The tram stopped on the river flats at Coalmont, just west of town where a large three story lower station, tension, coal bins and coal sorting tables were all built into the large wooden terminal. The tram crossed the flats to the base of the 'Anchor Hill'. There it ascended the twenty degree or so grade of the mountain for about half a mile. At the top of the hill was a large breakover station where the weight of the buckets was carried by steel rails in a twenty tower battery. The multiple towers were needed to spread the weight and strain of the cable arc over a larger distance at the abrupt angle change at the top of the hill.

After this the ground dropped away steeply and the tram passed over a deep ravine with the carriers and cables high in the air. The towers resumed where the ground rose again in a slight ridge. For the next mile the tram passed through gently ascending meadows, lodgepole pine and aspen forest. The towers interspersed with the forest in a not unharmonious manner. After a distance of rises and drops the ridge crested, and the tram started to descend to the upper terminal by #3 mine portal. It passed over easy ground, till it steepened at the lip of the canyon. Here was the mine operation, and upper coal bin making the tram $2^1/_2$ miles in length.

The twenty-seven towers were built on site with locally cut trees, the bents consisted not of squared lumber but of poles. They were all labouriously drilled and bolted together. Key sections were squared and fitted so that the joints would be tight. As with all trams the top beam carried the rocking cable saddles—they would be held to the beam by a large steel plate with an forged axle for the saddle. The cables were heavy, $1^5/_8$ inch diameter track cable that had periodic cable couplings. Cables with broken strands were removed and replaced and the couplings allowed sections to be changed. The track cables were too large and cumbersome to be handled in one piece. The west side of the tram was the heavy side and had slightly larger cable. Wrights Cables of Granville Island, BC, supplied the cable.

The traction cable was supported by 12-inch sheaves on each tower. They were mounted with babbitted bearings and oilers. If the traction cable flopped, it would be liable to derail itself thus cable guides were added. The tram was gravity operated, though an electric motor was needed to overcome the static friction at first for starting.[12] In an interesting feature, the coal buckets were specially built and doubled as underground coal tubs. They were designed to fit the mine wheels with a bracket. Mine railways were used around Blakeburn to draw the coal tubs with horses and hoists from underground. Electric locomotives then drew the trains of tubs to the upper tram terminal. At the upper tram station, the buckets were detached from their wheels and attached to the aerial tram hanger. There would be around 80 buckets on the tram and they were spaced some 400 feet apart to spread the load.[13] Again, timers were used to check the spacing at the terminal stations. A round trip on the tram took just over an hour.

The *Report of the Minister of Mines* goes on to explain the tramway construction.

> During the year 1920 there was a large staff of men employed in the erection of the surface plant, overhead tramway, and the installation of machinery and equipment for producing on a large scale. The tramway, which was started in November of 1920, is about 2 ½ miles in length and used for conveying the coal from the mine over the mountain to the tipple, the difference in elevation between the terminals is 1600 feet.[14]

At the tipple the coal travelled over screens to remove rock and pebbles. Also, coal sorters worked to break the large pieces and remove any rock. An inspector from the Canadian Pacific Railway oversaw the work when loads were being assembled for the railway.[15]

A large coal burning powerplant at Coalmont was built to supply electricity for the town and mine operation. The powerplant had a 600 horsepower Corliss steam engine coupled to 550 volt generator. Transformers upped to voltage for its transmission to the mine and camp via powerlines parallel to the aerial tramway

At the top of the tramway:

> a 50 HP AC motor generating set situated at the top terminal of the tramway connected by a belt and spur gearing to a 10 foot return Fowler grip pulley for controlling the tramway.[16]

Thus, the tramway used the weight of the descending coal buckets to generate electricity and act as a braking system.

However, the tramway was underbuilt when it was first constructed in 1920. "It was a complete failure and incapable of sustaining the load."[17] Cables broke when they were overstrained. Thus the mine shut down, due to its poor transportion, and the tram rebuilt in spring of 1921.

> During the year the 3/4 inch locked-coil stationary and carrying cable on the empty side of the tramway has been discarded and replaced by the $1^{1}/8$ inch locked-coil cable from the full side, this again replaced by a new $1^{5}/8$ inch track strand, reverse lay smooth coil, crucible steel cable, and the 3/4 inch endless or travelling cable had been replaced by a 7/8 inch "Lang" lay strand cable. The return grip pulley at the top terminal was fractured and was replaced by a Fowler grip pulley. The top terminal has been remodelled by the installation of a compressed air lifter which are used for lifting the box of the mine car by the frame by means of carrier arms to an overhead single rail connected to the tramway.[18]

Telephones, brakes, and electrical switch gear was also added to the top terminal.

At this time Anchor Hill was rebuilt and extended with more towers, and a maintenance camp with workshop and bunkhouse. It was possible to maintain the bucket hangers, their running wheels and rims on those wheels at Anchor Hill. The tensioning block purchase mechanism for the track cables was upgraded at the lower station while the lower station was reinforced with concrete.

The rebuild was a success. With the new cables the tram could haul around 450 tons per day (no length of day was given), and it would travel at about four miles an hour. The tram became a feature in the life of Blakeburn. And, as there were no coal storage facilities at the top of the mountain, only in the trains of buckets on wheels, the aerial tram was essential for the operation of the mine. The tram was also used to bring up food, supplies, and stores for the mines. If the tram was not working, then the mines would shut down. There was a fear of fire at Blakeburn and the wooden buildings, forests and coal would provide tremendous fuel which would suggest one reason for the lack of storage. The schoolboys in the district were hired to slash the tram right of way and keep the fire hazard down.

The tramway was a moving, working piece of machinery and thus prone to operational damage. Some telegraphed reports of this tram survive: "Runaway completely wrecked tower number four...endeavouring to work without tower but cannot yet say if this is possible." "Wreck this morning and trying another plan, if this fails we will have to close down."[19]

Cable failures caused delays in the mining operations. The track rope was replaced in 1921, and the manager stated "replaced by 31,000 feet of 7/8 inch of the same quality and make throughout and is at present working very satisfactorily."[20] The cables were replaced

A tram tower on the Blakeburn tram. Note the oversize wooden coal carriers.

to over a weekend by a crew to prevent loss of production time at the mine. This pattern is common in mines, logging operations and sawmills whenever they have a mechanical breakdown. It appears some of the runaway coal buckets on the tram were caused by unlocked cable grips on the buckets. If the grip was unlocked and uncinched, then the bucket was free to runaway on the track cable; to prevent this, and see that the automatic grip activator had moved the lever grip to the locked position, the operator would examine the grip on the outbound bucket to see if it was locked. If not, he would have to stop the tram when the bucket was at the nearest tower, walk out to the tower, climb it and re-lock the lever and grip.

There was a five-man crew on the tram each doing two hour shifts on the heavy side, light side, brakes, and weigh scale. In addition, there were two millwrights who maintained this tram. They rode the buckets once a day, to oil and inspect the cable and oil the tower sheaves. They also had the workshop at Anchor Hill where they looked after the bucket hangers and tower parts. The tower battery at Anchor Hill had also to be maintained with bolts, and rails to tighten. Apparently, the mechanics rode the buckets there in the morning, and dismounted at the battery. They would return in the evening. The millwrights were the only people allowed to travel on the tram. Pay was 75 cents an hour.

Young boys would try to ride the buckets for fun, though word would get around that it was happening and they were chastised. The moving parts, heavy buckets and heights were all prone to accidents and thus the ban on travellers. The mines at Blakeburn alone had, on average, one fatality a year.

The summer of 1930 changed all that. On a muggy, hot August 13 the skies were brewing for a thunderstorm. At 4 p.m. the afternoon shift went underground to start work. By 6:30 p.m. people were cooling off on their verandas when a large explosion was heard. Lumber was seen cascading through the air from a nearby mine portal. Blakeburn was a mine that was relatively free of coal damp, or explosive methane gas, and the mines were inspected daily for it as per Provincial Coal mine regulations.

Mine rescue teams were dispatched, and rescue calls went out to other coal mines in Fernie, Nanaimo, and Merritt. Rescue teams in Drager breathing lungs were unable to quickly pass through the wreckage and smoke to extract the shift. In the end 45 men were killed in the explosion and poisoning of the air of the mine. The cause of the blast was never determined. The lengthy inquest by the Minister of Mines questioned both the mechanical equipment and the climatic conditions. It is thought lightning may have hit the mine, or the lighting storm may have allowed gas and dust to escape from sealed off and disused workings. The rarefied and charged storm air complicated the issue. For such a small community, it was a horrendous tragedy.

The lower tram terminal, sorting station, and railway loading station at Coalmont.

Families were issued monies from a public subscribed relief fund. Workers Compensation also contributed in this time of economic depression. Due to the Depression, people got on with their lives and jobs, no matter how dangerous. Surprisingly, despite the modern machinery above ground, underground the work was done by muscle. The miners hand hewed the coal with picks, and shovelled it into the coal buckets. Brass tags were placed on the tubs to determine who had produced it as the miners laboured on a piecework system. The miners were paid $1.00 per ton, and on a good day could earn $20-25. Horses then hauled the buckets from the underground workings above ground. Miners also had to shore the pits with timber braces to prevent cave-ins.

Yet despite the timeless mining practices, life was also changing. The Great Northern Railroad converted its steam locomotives from burning coal to burning fuel oil in the mid-1930's and thus Blakeburn lost a good customer. Society at large was changing from being coal based to being oil based. Home heating systems converted from the bulky, and involved coal to trouble free oil. Transport patterns were changing from railroads to highways. The vast oil fields of Arabia were discovered in 1938. The writing was on the wall for underground coal mines. The accessible coal deposits at Blakeburn were also becoming worked out.

Blakeburn ceased operations in April 1940, and the mining equipment scrapped. Parts of the tramway, such as the traction cable and buckets were salvaged, the rest was left to the woodpeckers and wind.

Notes— 6

[1] Spokane. *Spokesman-Review*. November 27, 1899. p. 4.

[2] Ibid. p. 4.

[3] DM Wilson has written an excellent, online history of the area along Highway 3. (See *Greenwood and Phoenix* chapters. DM Wilson.) www.virtual-crowsnest.ca

[4] D. LeBourdais. *Metals and Men*. p. 62.

[5] *Report of the Minister of Mines*. 1915. p. K327.

[6] Ibid. p. K327.

[7] *Report of the Minister of Mines*. 1906. p. G265.

[8] Don Blake, *Blakeburn*. Wayside Press: Vernon. 1980. p. 10.

[9] At the turn of the century the thriving Slocan mining camps presented the Provincial Government in Victoria with a problem. To maintain sovereignty to the region, and not let the Kootenay be completely dominated by the Spokane money-men, reliable communication with the Coast needed to be established. To this end the Canadian Pacific Railway labored to build a "second railway" through southern BC; it would eventually come into existence as the Kettle Valley Railway. Vigorous competition, minerals, land grants, and difficult terrain all featured in that well elucidated story. The Vancouver, Victoria and Eastern Railroad, one of the shell companies owned by James Hill's Great Northern Railroad, was pushed through Coalmont from Princeton and joined their connections to Hedley and the US. It eventually became the joint trackage of the Kettle Valley for the GN and the CPR when the Coalmont Branch connected via Otter Lake, Brookmere, Cold River, Coquihalla and Hope to the BC Coast. Ibid. p. 13.

[10] Ibid. p. 13.

[11] The Blakeburn aerial tram holds an interesting place in this story. One the one hand it is one of the better documented trams, with construction details and photographs, on the other hand key details are missing. First, in the pantheon of aerial trams in BC, Blakeburn is probably second only to the Premier Mine in Stewart. The Coalmont tram entered into the popular imagination and did not lapse from the collective memory as so many others did. Case in point-a photograph of the Coalmont tram in a Princeton restaurant piqued the author's interest in aerial trams several

years ago. Second, the tram is a more recent construction, it operated until living memory, and had some records survive. That said, the author has had innumerable problems deducing the builder of this tram.

The author went to great lengths to try to solve this problem. The normally encyclopaedic *Report of the Minister of Mines* is mute on the issue. Don Blake describes the tram in detail in his book *Blakeburn,* but omits to mention the name of the builder. Communication with Don Blake concluded that he is unaware of the tram's builder. Attempts to locate Coalmont Colliery Company records or sift data from the papers at the Princeton District Archives all came up with nothing. There is one tantalizing fact about this tram and it is cast into the tram gripwheel on display at Coalmont—that is the phrase "Leschen Company." Now, given the involvement of Riblet with Leschen, and as we have seen, the overlap of their work (including both claiming paternity status to the Britannia Beach tram), it is possible that this tram was a Riblet-Leschen joint effort. If this tram was constructed by Leschen, why, in their extensive and epistolary essay on the wonders of the Leschen Tramway in their 1930 publication, did they not mention their involvement with Coalmont? It could be that the tram was a joint effort with stock parts ordered from Leschen, the shops at Blakeburn, and hired contractors. However, the poor performance of this tramway on commissioning suggests some amateurish engineering so the builder's name was omitted from all records. In the end, the lack of records only allows us to conjecture on the topic.

[12] Ibid. p. 24.

[13] Ibid. p. 25.

[14] *Report of the Minister of Mines.* 1922. p. G322.

[15] Both the Canadian Pacific Railway and the Great Northern Railroad were large users of coal; in the 1920's their steam locomotives burned coal. An inspector was at Blakeburn to make sure the quality of the load was good. If there was too much rock, the load would be rejected. Too much rock in a load of coal would cause trouble for the locomotive firemen. The silica in the rock would fuse with other rock to make large clinkers on the grates in the fire box and prevent the passage of air for proper combustion of the fires. Also, the clinkers would form with the grate to cause a cool spot in the firebox, which would reduce the heat content, and thus the capability of the fire.

[16] Ibid. p. G322. Precise details on a Fowler grip are unclear. It probably is a variant of the Hallidie grip, and was redesigned and improved by the resident engineer of the tram company in the same manner engine builders named the Vauclain smokebox, or Baker Valve Gear after their inventor.

[17] IBlake Op cit. p. 35.

[18] Report p. g322.

[19] Office Correspondence from Coalmont Colleries. Cited in Don Blake's. *Blakeburn*. Op. Cit. p. 25.

[20] There is a tenuous connection with Blakeburn to the Riblet Tramway Company. While the builder of the Blakeburn tram is expressly never mentioned, neither in Don Blake's book, nor in the *Report of the Minister of Mines*, it is suspected that the tram was built by the coal company. In Royal Riblet's 1921 Diary there is mention of him visiting Coalmont on company business, though the citation is not elaborated upon. Royal Riblets travels are listed in his diary. Royal Riblet Papers. 1921 Diary. Cheney Cowles Museum. For Tram details (see Blake. Op. Cit. p. 25.)

People and Farther Places

24. Royal Newton Riblet

Royal Riblet was the younger brother of Byron, and he had a supporting role in the tram building business. In many respects, Royal even overshadowed his brother, whose business, talents and enterprise were often mistakenly attributed to Royal. Born in 1871 he grew up in Osage, Iowa. There he helped around the family farm, and like most youths, learned to tinker with things mechanical. To this end, he became passionate about bicycles. This being the first age of bicycles, when they were popular in the 1890's, he rode them everywhere and even competed and won the state bicycling championships.

Perhaps it was on the bicycle that Royal Riblet became determined not to let the grass grow under his feet. His peregrinations started early, and would later continue when he worked for Byron's tram business. At the end of the nineteenth century he moved to Aberdeen, South Dakota where he opened a bicycle shop. In turn he met his first love, Mary Gertrude Knapp, to whom he penned some debatable poetry. Eventually, they married and had two daughters and one son (Virginia, Marjorie and Jackson).[1]

By the turn of the twentieth century, Byron Riblet had established himself in Sandon, the West Kootenay and Nelson building tramways. To assist in the business, he invited his family north.[2] Consequently, Byron's parents, and brothers settled in the beautiful town of Nelson. Sadly, such togetherness was not to last. The father William Jackson Riblet died in 1903. And tragically, Mary Gertrude succumbed to illness in the same year too. They were buried at Nelson, BC.

Beyond the family crises, the tram business boomed. Trams were ordered all over the Kootenay, Boundary country, and orders came in that year for Vancouver Island and the Coast. Royal Riblet acted as a salesmen, assistant engineer, and agent for the company. If a mine wanted a tramway, a survey and profile would be made. The topographic surveys were then sent to Byron who would approve and design tower locations, and terminals. Then Royal would return to Nelson where he would oversee the fabrication of the metal parts in the blacksmith shop. The ore buckets were forged, as were tower brackets, bucket hangers, station rails, and station stanchions. After the parts were made and freighted, someone had to oversee their installation. Often, this task was performed by Royal Riblet. This is how he travelled extensively on business. He was the agent on the Conrad, Yukon trams, and he travelled south to South America on business too. Royal worked for Byron for over thirty years. As a show of affection for that loyalty, Byron transferred a large sum of money to Royal. With it Royal set about building a legacy that lasts to this day- the Eagle's Nest Estate near Spokane.

Situated on a high volcanic bluff, overlooking the Spokane River, and surrounded by Ponderosa Pines, the house and location are striking. The house was built in 1924 to a Florentine Style by local architect George Keith.[3] A three story structure which was appointed with many modern electric conveniences, heat, food and refrigeration. While the grounds were tastefully filled with a swimming pool, rock garden, chess board, "as well as bridle paths, ski slope...croquet court [and] skating rink."[4] Offsetting the recreation was a rose and herb garden and lily pond. In the five years after it was built some seven thousand people came to view it and Royal's patience. It was considered the most modern house in the US when it was built, and so influential that Twentieth Century Fox studios even made a newsreel about the house.

Mechanical aides were built in and reflected the worldview of its owner. The medieval style gate-house, had a cable gate closing mechanism which allowed the barriers to be activated from the garage while a miniature tram retrieved the mail.

Royal lived at the house until his death in 1960. He used the estate as a place to think and to tinker. One idea of his was a tank tread to go over a tractor wheel. It would increase the bearing surface exposed to the ground and allow for greater traction. Royal devised it about 1918, though he refined it over the next twenty years. The US Army conducted tests on the wheel in 1943 and were interested in applying it to war work, though, in the end, the technology transfer did not happen.

Riblet devised other items too: traffic indicators for cars, ore bucket swivels, drill bits, aerodynamics for cars, swing staging and automatic parking garage.[5] Royal was also busy with the tramway business. He assisted his brother until 1933 when they had a falling out over money. As a result, the two had an irreconcilable break. Royal then went on to found his own tramway business—the Riblet Airline Tramway Company. Needless to say, customers often mistook the two companies. Royal appears to have built only two complete tramways–one at Bishop, California and one at Algoma, Ontario. His draughting drawer index list plans or rebuilds for other trams. Some of the drawings have survived. The index intriguingly cites locations such as: Grand Coulee Dam, Hedley Mascot, Fernie coal, Weyerhaeuser Lumber, Ruby Dam, Gold Fields (Zeballos), Braden Copper, Hollyburn Ridge, Tin Mine (China), Panama Canal, Zincton,

Sheep Creek, Last Chance Mine, Washington Iron Mines, Philippines, and Palm Springs.[6] Not all the drawings have survived, but they do indicate the diverse installation and contracts Royal had in the twenty five year period after he parted from Byron.

The preponderance of records in favour of Royal over Byron is astonishing. This has assisted Royal in eclipsing his brother's shining works. Even the citizens of the fine city of Spokane are oft to confuse the two brothers. Compounding the issue was that Byron Riblet has left so few records. Secondary mention of Byron in Royal's documents is limited too.

Royal was busy in other areas of life as well. He had eight wives over the course of his life though most of these unions ended in divorce but a higher net worth for Royal. Between the marriages, trams, and estate, he continued to tinker. He had a workshop looking out from the bluff where he liked to devise.

Royal built a sight seeing tram up from a paper company, immediately adjacent to his house across the Spokane River, over the river and up to a crevice in the basalt bluff. Using standard bicable tram designs, Royal constructed a single reversible tram. For the carrier, he used a car chassis, engine and drive. A hole in the differential casing allowed a large pinion to be driven from the car's rear end. A large cable sheave was then bolted to this pinion, which allowed the motor to control the traction cable. That cable was spooled around the large sheave and out through guide sheaves. In this way, by driving the car, one could control the movement of the carrier—both speed, direction and braking capacity. The tram operated until 1945 when the cement dust pollution from the nearby plant was wearing on the cables and the system deemed unsafe. The cables were taken down in 1956 and stations were dismantled. The bleached remains of the carrier lie forlornly at the Eagles Nest Estate.

That said, the present owners of the Eagle's Nest Estate have done an amazing job in converting the estate into the Arbor Crest winery, park and tasting house. They also have spearheaded research into the Riblet family by donating papers to the Cheney Cowles Museum, restored the Eagles Nest Estate, and promoted historical publications. It is a very business-like approach to the legacy of an eccentric, yet exceptional, individual.

25. Encampment, Wyoming

Atop the Continental Divide in southern Wyoming lay important mineral resources. Under the aptly named Carbon County were rich coal and copper deposits. An Englishman named Ed Haggarty located these copper outcroppings in the Sierra Madre mountains in 1897. The ore content was an astonishing 33% copper at the mine. In time he contracted with George Ferris, Robert Deal, and James Rumsey to develop the property.[7] Using the first two letters of the men, a name was coined: the Rudefeha Mine. Later it became the equally cumbersome Ferris-Haggarty Copper Mine. Of course development work charged ahead with stripping of the overburden, building roads and mine buildings. Shafts were driven and the rich ore removed.

As in the development pattern of the West, support towns sprang up and merchants sold their wares to cash in on the boom. In this way, the town of Grand Encampment was started by the copper company; it was situated roughly eighty miles west of Cheyenne. Part of the development included a copper smelter. The territory was rugged with the mine situated high on the mountainous divide, in timber, and with no local railway. The winters were very extreme. A nearby gold strike, provided further impetus for the region.

To foster improvements the Boston-Wyoming Smelter and Power Company was formed to oversee industrialization. Their smelter had a capacity of one hundred tons of ore per day and was finished in 1902.[8] Transport was a problem as both the ore and coal had to be moved by horse wagons. To eliminate one of these problems, a sixteen mile aerial tramway was built.

The contract for the tramway was given to the Leschen Tramway Company of Saint Louis. Byron Riblet was the Chief Engineer on the project. Riblet had built eighteen tramways in BC by the beginning of 1902. That year saw armies of men arrive to build the gigantic system. Carpenters came from Canada, and available locals were also hired. Tent construction camps were raised to house the multitudes.

> The construction of the aerial tramway progresses nicely, towers for the section of four miles having been completed, and the second section commenced. A new camp will soon be established at a point about half way between the smelter and the Ferris-Haggarty. The big cables have arrived at Walcott, and are being loaded this week on the transportation company's wagons. The cables come on spools weighing seven tons each. The Transportation Company has three new wagons which are guaranteed to carry 15 tons each,

and these wagons will transport the cables 50 miles to towers. Fourteen horse teams will draw the loads. The supply of timbers for the towers is not quite equal to the demand, owing to the bad roads in the hills. The line of towers as they stretch from townsite across the sagebrush over the foothills present a very fascinating appearance, bespeaking a fulfillment of the hope in the future of Encampment and its smelter.[9]

With some 150 men on the job, the work progressed quickly. By May of 1902 a nine mile length of towers were up, and the first cables were being raised.

Of course, before the age of bulldozers, diesel tractor trailers, and hydraulic excavators, moving the large spools of cable provided its own set of problems. There were difficulties in "handling of the cables which are shipped on spools weighing from $5\,1/2$ to 7 tons each. The spools are hauled in from the railhead by ten horses, and in taking them up the line as many as fourteen horses are required to make the pull."[10]

The historian Moulton then elaborates on the tribulations of tramways with excerpts from period newspapers:

Several of these large spools have been in the tramway yards in town this week, in all five spools weighing $35\,1/2$ tons and containing 32,700 feet of cable. Great difficulty is met in unloading the spools from the wagons as they come from the railroads and reloading when ready to be taken up the line. A wagon bearing a spool is driven alongside the skidway and anchored, and six horses are hitched to the spool to land it on the skidway while the 'snubbing' process is going on the opposite side. When a spool weighing six tons gets started rolling it is not easily stopped, and several bad accidents have narrowly been averted while unloading. Two spools now in the yards have broken through four 10 x 10 timbers on which they were placed when first unloaded from the wagons.[11]

There were 33 miles of $1\,1/4$ inch track cables. Cables could be spliced together as needed.

Construction continued throughout the summer. There was much to be done: the tramway had 370 wooden towers and two terminals. The line also was broken into three sections, with two intermediate power stations. Steam engines, and their ancillary coal burning boilers, moved the traction ropes from a full stop. Sixteen tension stations allowed for thermal and load variances in the cable.

The tramway took over a year to build and was not in operation

Riblet towers on the treeless plain of Encampment, Wyoming

until June of 1903. By which time its 985 thousand pound capacity ore buckets moved in unison. Their route would cover two miles east, ascending the Rocky Mountain Crest, then for a fourteen mile descent in to Grand Encampment. The tram was a success and had cost $405,000. When completed it was the longest system in North America and would have retained that title until 1926 had the Wyoming tram continued to run until that date.

Another intriguing fact about the Ferris-Haggerty tramway was that its right of way in the trees provided a cache point for Butch Cassidy's criminal gang. One of the gang members, Ben Ellis, had worked in the area and knew it well. Encampment was even listed for the site of a possible holdup, with its sizable payroll acting as a magnet. Gangmembers hid out in the nearby Browns Park after their raids; they also journeyed to their Hole in the Wall hideout. In the process, they stashed some purloined mail pouches on the tram line "near Bridger Peak."[12] The pack then split up as they were being pursued by an irate, armed posse. Of course this is before Butch Cassidyand the Sundance Kid allegedly finished in a Bolivian oblivion.

There is a school of thought which thinks that Butch Cassidy survived the Bolivian escapade, went to Europe, had adventures in Mexico, married, contacted his younger sister and settled in Spokane. Over the years historians, forensic pathologists, and other Old West types have studied the subject closely. (This is where the coincidences cut close to home.) By his time in the Northwest, Butch had changed his name to William Phillips and started a self-named engineering firm in Spokane. After it failed to pay, it is said—that Butch Cassidy worked for the Riblet Tramway Company sometime in the teens and after his supposed death.[13] Without company payroll records or photographs to verify this claim, documents that have not surfaced in the course of this study, this claim can only stand as conjecture. Nonetheless, such nuances make for interesting reading.

In the process of all this activity the Ferris–Haggerty Company in Wyoming sold its interest in the Encampment Mines to the North

Over the crest of the Rocky Mountains with a sixteen mile Riblet Tramway. Wyoming 1904.

American Copper Company for one million dollars. With the purchase came interests in the mine, town, water, electricity, tramway and smelter. One major problem was the lack of railway facilities. Supplies and materials had to be hauled the 44 miles to the railhead at Walcott.

It was difficult country to operate in. In the winter, snow fell to a depth of twenty feet, a fact that caused difficulty and danger. An assistant engineer on the tramline named Antoine LeMieux set out in the winter of 1903 to fix a downed telephone line on the tram. Snowshoeing over the deep snow he and his partner were overdue in arriving back. Fearing danger, the locals set forth search parties to look for the snow tracks. The pair's bodies were found amid the slides of Cow Creek Canyon.

There were devastating slides in the business world too. The North American Copper Company sold its ownership of the operation to the Penn–Wyoming Copper Company at the end of 1904. By which time the price of copper had slumped. It seems that the former company's owners had quickly sold out to evade bankruptcy. The firm's debt to income ratio was high, and shareholders of the previous company were left to pick up the bill. Share prices slid, and the North American company remaining assets were sold at a fraction of their cost. Of course cheated shareholders headed to the courts with the Penn–Wyoming Company in tow. And to cap it off, the ore body at the Ferris Haggarty was shrinking in size.

In the short term the problems were alleviated, and the mine continued to work for the next three years. Low copper prices and high freight costs made the operation marginal. Penn–Wyoming managed to elevate the size of the stock issue to bring in more capital. The Encampment Tramway Company's share issue went from 40,000 to 450,000 stocks in circulation. It looked as if the project had overcome its difficulties. Even a railway spur was started but would take several years to be operational.

Legal wrangling continued unabated. The two companies fought over the payment schedule and stock transfer valuation. The mine worked until May of 1908, when metal prices and a cash crunch terminated the viability of the Penn-Wyoming Company. Cussedly, the Union Pacific railway spur arrived in August of 1908 too late to be of use. The year 1909 saw a Virginian group under A. Hawse buy out the board and optimistically take control of the company. Alas, the company's debts and cash flow pushed it over into receivership in early 1910. More lawsuits were filed and thus the assets were tied up in long courtcases as their worth depreciated to nothing.

Snowfall demolished the buildings and water filled the mines. The tramway and smelter were abandoned. The town of Encampment went into decline, with desultory lumbering in the nearby national forest providing some employment. Today, parts of the tramway and two towers have been preserved in the Grand Encampment museum. All in all, it was a sad denouement for a record breaking cableway.

Notes—7

1. Barbara Loste. *"Eagles Nest—Home and Workshop of Royal Riblet."* Arbor Crest Cellars. 1995.

2. Ms. Loste creates an off-colour malapropism with a spelling error. She states that Royal "took up residence in Nelson, British Colombia." Considering the recent emanations from Marc Emery, Jean Chretien, and the US Drug Enforcement Agency about marijuana, the Criminal Code of Canada and trafficking in that fair Canadian Province, the joke is quite succinct. Also, a present-day promenade along the streets of Nelson reinforces this idea as pot addled panhandlers persistently pester pedestrians there. Quotation from Loste. Ibid. p. 8.

3. Spokane has many fine homes; many were built from the proceeds of Kootenay and Coeur d'Alene Mines. Ibid. p. 3.

4. Ibid. p. 4.

5. Ibid. p. 17.

6. Royal Riblet Archives. Cheney Cowles Museum Spokane. L 94-50.

7. Candy Moulton. *The Grand Encampment—Settling the High Plain.* High Plains Press. Glendo, Wyoming. 1997.

8. Ibid. p. 91.

9. April 4th 1902. *Grand Encampment Herald.* As quoted in Moulton. *The Grand Encampment.* p. 95.

10. Ibid. p. 94. Quoted from *Encampment Herald* June 20, 1902.

11. Ibid. p. 97.

13. Ibid. p. 136.

14. This was a most peculiar research item and one that was unexpected. Butch Cassidy, in this version of the story, later went back to Wyoming to find some of the cached money, but he died relatively poor in 1937. (See www.utah.com\oldwest\Butch_Cassidy).

The North: Alaska and the Yukon in the Edwardian Era

26. Conrad City, Yukon

The rapid expansion of mining in Canada's North was fostered by the Yukon Gold Rush. Thousands of stampeders rushed to the placer creeks around Dawson for a chance at easy money. Most of the stampeders arrived too late and were forced to work for wages while deciding what to do next. Fortunately, the North is abundant. Other prospects turned up. These included new gold rushes at Nome, Alaska, Atlin, BC and Conrad, Yukon. As a result, prospectors fanned out to search nearby areas in the North.

The mines at Conrad on the Windy Arm of Tagish Lake were located and staked in 1899. William Young named the Montana claim after his home state. John Pooley and Jack Steward located and claimed the Mountain Hero. They later registered the Venus claim as well. The mines were located close to the BC-Yukon border, and until surveyors accurately determined the setting, there was some disputeover which domain it was in. Tagish Lake, Bennett Lake, Marsh Lake and Atlin Lake all form an interconnected chain which allows transport by boat and sled in winter. Steamboats plied the lakes and connected Atlin to the White Pass and Yukon Railway at Carcross in the Yukon.[1]

The remoteness of the Windy Arm site, and the elevated location of the vein hampered speedy production from these mines. The mines were above the tree line in the exposed peaks of the mountains. Food and supplies all had to be packed in. More development work progressed and some ore was hand sacked and transported by early 1902.[2]

Now a charismatic man entered the story. John Conrad was a larger than life businessman and promoter. Originally from the US South, the disruptions caused by the Civil War forced the family west. In the process they settled and worked in Montana as 'whiskey traders' bootlegging liquor into Canada during the pioneering 1880's. John Conrad speculated in the cattle booms. Life was going well and the family attained respectability when the family was embroiled in a murder involving a suspicious poisoning, inheritance money, a dubious doctor and relatives. Later, the Crisis in 1893 left him in further dire straits. Consequently, he wrapped up his business in Montana and made investments in the Copper Boom on Prince of Wales Island, near Ketchikan.[3]

After scouting the north, liaising with the shipowner John Irving, and searching for investment opportunities, Conrad optioned the mines at Windy Arm in 1903.

And in course, the Conrad Consolidated Mines Incorporated was formed.[4] It was to become the driving force behind the community.

At the mines, hand work was progressing with the driving of tunnels, sacking ore, and exploring the orebody. Sixty men were labouring to drive the tunnels hundreds of feet and extract the ore. By the end of the year, 2900 feet of tunnel had been dug. Conrad claimed it was "one of the world's greatest mines." Between the hyperbole and the hucksterism, prospects for the Conrad Mines looked good.

In order to improve transport from Bennett Lake to the 3400 foot level that the Mountain Hero mine was situated at, an aerial tram was organized. Royal Riblet was contacted in July 1905 and a tramway ordered for the Mountain Hero Mine. That tram cost $80,000 and ran 18,600 feet in distance. The vertical fall was 3400 feet, and the track cable diameter was $1\ 1/8$ cable.[5] Royal Riblet was then retained to do a survey for an additional tram.

Materials for the first tram were shipped in from Seattle—via Skagway, the White Pass and Yukon Railway, and the lake steamer *Gleaner*. Then "the supplies and equipment" were hauled up the mountain using "45 mules and a dozen pack horses."[6] Through the end of summer and into fall, crews laboured to build the tramway before winter. R. Lanyon, the Nelson contractor who did much work for Riblet, was the chief contractor, and he used fifty men to assemble the tower timbers, affix sheaves and saddles, and build terminal stations. In September of 1905, the 90,000 feet of steel cable for the track and traction ropes arrived.

During the stretching of the track and traction cables, a capstan broke. These capstans were similar to ship's anchor winches, and they allowed the construction crews to tension the heavy cables. Spare parts for the capstan were ordered from Seattle, though, it being so late in the season, it dealt a blow to the timely completion of the tram. Fortunately, a break in the weather facilitated the week long process of lifting the cables.[7]

Other crews built a pack road to the mine and a connecting road to Carcross. The mines needed buildings as well and which were built of stone. Ore bins, bridges, hotels and wharfs were also constructed. To move in the materials to Carcross, trains were chartered on the White Pass and Yukon Railway.

Meanwhile, John Conrad was expounding on the quality and extent of ore at the Montana mines. This he did to newspapers, government officials, and fellow businessmen. Assay reports noted that the ore contained silver ranging around 1000 ounces per ton. The ore was labouriously sacked and shipped to Tacoma for smelting. To offset the problems of shipping unrefined ore, a smelter was considered for the region. But the dearth of coal in the North, the distance in its haulage and financing delayed this idea.

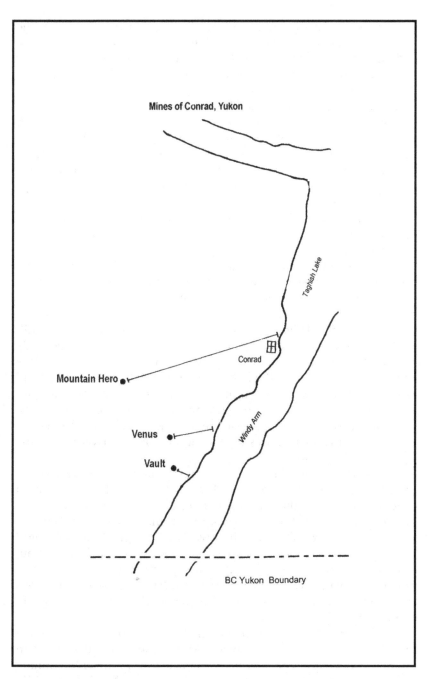

Map of Conrad

Conrad and his partners had purchased 160 acres of land on the lakefront to assist development in the mines. In fact, as with many early mining communities, the land speculation returns were similar to mining and an important source of revenue.[4] Thus the town of Conrad was platted (surveyed). At first the metropolis was a tent city, though later log cabins and hotels were erected.

Later, a semblance of civilization was realized because along with the hotels and houses of joy, came the police, religious ministers, and nursing sisters. Houses, restaurants, and print shop also settled in Conrad. News could come in via a telegraph line which was also rigged up to connect with the Yukon telegraph and the outside world. A tote road to Carcross was also built.

Additionally, the Venus mine was being developed. Miners were at work, and Riblet completed an 1800 foot tram to the beach. That tram cost $5300.[8]

The Mountain Hero tramway, like the period of its construction, continued being difficult. It was quite a long tram way, and its 22 towers undulated up from the lake and over hills to the mine. But "the practical limits on tramway length and tower spacing apparently had been exceeded."[9] There were many mechanical failures as a result. Underground, the Montana vein had narrowed and the mother lode had not been found. Thus work on the mine stopped in 1906. This impacted the tramway, for while it was still used to haul supplies and ore from surrounding mines, it did not have a ready source of weight. The tramway was gravity operated and thus "in order to work the tramway, to keep several men shovelling loose rock from the mountainside into the buckets, this rock being dumped afterwards near the lower end of the tramway."[10]

Meanwhile, down the lake the Vault mine was also developed. Bunkhouses were put up and adits were driven. It was quite an exposed camp with the mines situated in a canyon with rock walls. The bunkhouse was built against a rock wall. A small tramway was built for transport. "The Vault tramway was designed for fairly light loads using the half inch cable that had been used to haul in the main cable on the Montana [mountain] tramway." The tram used a gasoline motor to move the cable. Thirty men were at work clearing 150 tons of ore which was shipped to the Howe Sound Company smelter in Crofton. (See Mt. Sicker. q.v.)

William Mackenzie was the Eastern Capitalist primarily investing in railways. Together with Donald Mann, they were interested in building a railway to the Yukon. Mackenzie also speculated in mining ventures including ones at Conrad. In this way MacKenzie hired

Joseph Tyrrell to investigate the mining claims there. Tyrrell had worked for the Geological Survey of Canada, and made important discoveries in Alberta.

Thus in late 1906, Tyrrell was encamped in Conrad. There he investigated the mines, examined the operations and took photographs. In November of 1906, Tyrrell wrote a secret, damning report on the minerals and prospects of the mines at Conrad. Phrases such as 'valueless' and 'of no value' poured cold water onto the previously recorded bluster of John Conrad. MacKenzie heeded Tyrrell's advice and quietly pulled his money. Tyrrell went on and also criticized the overinvestment there and the tramway which "stood as a monument of erroneous judgement and reckless and wasteful extravagance."[11]

Astonishingly, as Tyrrell's report was secret, Conrad still blundered forward seeking investments and buying machinery. Though the winter of 1906-7 cooled activity in the mines and when the spring thaw arrived people's ardour with the region had waned. Also, 1907 was the year the business cycle burst. Caution was in the wind and settlers moved from Conrad to other places in the North or down south.

By the summer of 1907, John Conrad had purportedly secured millions in capital from the East and Europe and was embarking on building a smelter in the North. Using part of that capital, Conrad did build a modern concentrator at the Venus tram terminal. The concentrator cost $60,000. Yet despite the modern machinery, it had a difficult time removing the rock from the ore. The tables, tanks and vanners had to be altered numerous times to try to maximize the separation. When done, it was with limited success.

With the future of Conrad City in doubt, John Conrad then looked around the North in search of other investment opportunities. Oddly, in 1908 he even contracted Royal Riblet to conduct a survey over the Chilkoot Pass where the three other aerial trams had worked some nine years earlier. Despite their failure to compete with the White Pass Railway, and that railway's physical removal of the trams from that pass, Conrad persisted. "A total of nearly 30 miles of double tramway was planned"[12] to connect with lake steamers over the Chilkoot Pass, Lindeman Lake, and Bennett Lake. In the end the tram was never built.

By 1909, the Venus mine was still producing, but the town's inhabitants "had dropped to twelve."[13] By the next year the town's post office closed, marking the end of the adventure.

27. Akun Island, Alaska

On Akun Island there is a large volcanic fumarole. These volcanic ventholes collect native sulphur over time, a phenomenon that is common to the Pacific Ring of Fire. The Makushin Volcano—two islands west of Akun at Unalaska—has similar deposits. About the time of Shackleton's adventures in the South Atlantic in the First World War, a Chicago entrepreneur wanted to mine the Akun sulphur and sell it to industry who would convert it into oil of vitriol.

Byron Riblet explained problems with the Akun tramway construction to a newspaper reporter of the day.

> Tom Graham, my General Superintendent, was in charge of the work and they were frozen in and forced to stay all winter. The tramway ran from an old crater down to the beach and was to be used in retrieving sulphur.
>
> The project was promoted by a Chicago man. He spent about a $1,000,000 on it. During the long winter on this bleak, barren island on which nothing grew of any kind and on which the only life was a few foxes, the cook for the five other men there went crazy from the stark, staring lonesomeness and moonshine that he distilled on an improvised still. He died before spring and with dynamite and much effort, they managed to hew a grave out of the frozen ground on the beach.
>
> In the spring the Chicago promoter visited the island and was told of the death and he exclaimed: "This is were the townsite is to be and that grave lies right in the business district."[14]

The mine was a total failure, and the tramway was finally demolished by volcanic activity in 1921.

28. Ketchikan District

At the southern tip of the Alaskan Panhandle is the Port of Ketchikan. Now known as a cruise ship "local colour" whistle stop, Ketchikan also has a large pulp mill and forest industry in the overcut Tongass National forest. Despite this activity it is generally not well known that the district had an extensive mining history. Prince of Wales Island lies across the channel from Ketchikan and is one of the larger landmasses in the Alaskan archipelago. On the island lay the source of timber and minerals.

During the many nineteenth century mineral rushes, prospectors fanned out in search of wealth. It seems that an Englishman Captain Crooks first explored the area in Kasaan Bay on Prince of Wales Island in 1870, about the time of the Stikine gold rush. On the beach he found showings of copper. A generation later, the Klondike Rush moved mining attention elsewhere. Yet the Alaskan Panhandle was not forgotten. In time, Crooks told other prospectors who staked the area around the turn of the century.

By that time mineral production on Prince of Wales Island was charging ahead. In the mid nought years, there were almost fifty active mines. One was the Mamie claim on Kasaan Bay—it was discovered in 1902 by C. Fickert and produced ore by 1904.[15] New York money was brought in and a smelter erected at Hadley, on Kasaan Bay; the complex was equipped with a 5500 foot aerial tram to from the Mamie Mine to the smelter. "In 1905, 12,600 tons were shipped, and in 1907, 30,000 tons. Average values of the ore were $3\,^1/_2$% copper and 0.05 oz in gold."[16] The Granby Consolidated Mining and Smelting Company bought the mine in 1913 and it is to Granby in Alaska that Riblet sold an aerial tramway. Granby also operated the nearby Rich Hill mine as well.

Copper was also located on the Western slopes of the Island and these mines began production. Appropriately named hamlets of Copper City and Coppermount sprang up on Cordova Bay. Near Coppermount was the Jumbo Copper mine and it was just one of many on the bay. The mine was located at 1700 feet height on Mt. Jumbo. It was discovered by A. Shellhouse in 1897, and was producing five years later. It was tough country to work in for more than one miner died in marine, gun or hiking accidents.[17]

Charles Sulzer bought the mine and became a very respected mine operator. Through his efforts a Riblet aerial tram, ore bunkers and wharf were built. From the mine the ore travelled 8250 feet over the tram to the wharf, where it was shipped to smelters at Ladysmith,

Crofton or Tacoma.[18] Over the years the chalcopyrite deposit paid handsomely, for the Jumbo was a stable employer in the region until it shut down in 1920.

29. Prince William Sound, Alaska

On the oleaginous ocean reaches of Prince William Sound, Alaska, there were some early mining prospects. The most famous was probably the Beatson copper mine at Latouche Island. Latouche and Knight Islands are within the Chugach National Forest on the western portion of Prince William Sound.[19] Kennecott Copper owned the large underground and pit operation there; they built a wharf, concentrator and outbuildings on Latouche Island to support the mine just after the turn of the century. By this time the Guggenheims had ownership of the Kennecott Copper Company which may explain the large investment and extensive undertakings at the Beatson Mine. Also in the lee of the Kenai Peninsula was another prospect—the Jonesy Group.

Another indented coastal feature on Prince William Sound, just north of Latouche Island, is Knight Island. In the centre of that Island is Drier Bay, a deep, protected inlet. In turn, a Klondike-era prospector named Hemple developed an interest in the Knight Island copper claims. Atop the hill, at 1000 feet, was the Jonesy adit. By 1907 "considerable development work, including the erection of a wharf, offices, steam plant, and aerial tram, was done on the property."[20] The owner of the operation was the Knights Island Consolidated Copper Company.

The Riblet Tramway Company was the builder of the tram which stretched up from Drier Bay up the barren hill on wooden poles. Little is known about the mine or the tram other than the upper tram terminal was damaged by snowslides in 1908. All told, some 330 tons of 3% grade chalcopyrite ore were mined, though production had ceased by the time the 1918 US Geological Report was written.[21] It appears the mine was abandoned and much of the equipment was salvaged. Today the tram cables still snake their way from the beach up the slope. The harsh weather and heavy snow has deteriorated what remains of the mines of Drier Bay.[22] In addition, the Valdez region was greatly affected by the 1964 earthquake when many structures and landforms collapsed during that record shake.

30. Dolly Varden

Near Anyox, on Observatory Inlet, eighty miles north of Prince Rupert, BC, is the Kitsault River. Interest in the region had blossomed at the turn of the twentieth century for the Granby Consolidated Mining and Smelting Company looked for other jewels to put into its crown after success at Greenwood. They developed the large mines and smelter at Anyox in 1914. Anyox had a large deposit of low grade copper ore.

Nearby, Ole Evindson searched the Kitsault River region; it was about twenty miles from Anyox. There he staked the Dolly Varden mine in 1915. It was in very rugged terrain, and also exceedingly wet, being situated in the exposed north coast forest. Despite this, the silver vein was of such high content that the mine started development immediately even in the face of a shortage of materials due to the war. A narrow gauge railway was built stretching fifteen miles up that river. A new Heisler locomotive was ordered from the factory and used to navigate the steep, winding way in the primordial forest. In the process of building structures the original operator went bust as he underestimated the costs involved in such a remote location.

Riblet built a medium sized tram and ore bins to bring the ore from vein to the narrow gauge loading track. The bins and towers were built from forest cut lumber and logs. Another ore bin was built at tidewater and the ore was barged down the inlet to Anyox. The mine produced 1.3 million ounces of silver, but had a very short life and shut down in 1921.[23] In the oracles of the obscure, the Dolly Varden mine ranks highly.

Apparently, the Kitsault region sprang again to life in the 1970's. The old claims were reworked and new machinery, buildings and investment were showered onto the area. In fact, there is a 'new', extant ghost town with the houses, stores and facilities that were built at this time and now standing idle. Thus in the legacy of post-war resource towns in BC—towns such as Tumbler Ridge, Tahsis, and Cassiar—here is one more to add to the list.

Anyox was a big operation with mines, railways, dams, and townsite. Anyox had a Riblet tram too. It was two miles long and worked at the Bonanza mine, a few thousand feet from the smelter. Anyox was in full operation during the First World War. It expanded its smelting plant in 1923 to profit from the lower sulphur content copper ores available.[24] By the mid-decade the mill was churning through 4800 tons of ore per day. Unfortunately, the collapse of the copper market in the Depression finished the prospects for Anyox and it was closed forever. Anyox has been reduced to foundations and to a mere memory in but a few minds.

The lower station of the Riblet tramway at the Dolly Varden mine; Kitsault River, BC. Photo: Vancouver Public Library No 2253.

Notes—8

1. John Irving was a famous steamboat captain. His father was a pioneer in Portland, Oregon and a famous riverboatman himself. Irving *(Pére)* freighted for the BC Gold Rush and was in competition with "Bully" Wright. John Irving expanded to BC and the Yukon and his boats plied the Stikine, Skeena and Yukon Rivers, Atlin and Bennett Lakes. The family house in New Westminster, BC, is a museum.

2. Murray Lundberg. *Fractured Veins and Broken Dreams*. Pathfinder Press. Whitehorse YK. 1997.

3. Apparently, there were many mines on that wet and cold Island. These mines boomed for a period of ten years from about 1897-1907 and are now all but forgotten. The Treadwell Mine in Juneau was more famed. After this period, capital went in search of timber and fish.

4. Lundberg. Op. Cit. p. 25.

5. Ibid. p. 28.

6. Ibid. p. 29.

7. Ibid. p. 43.

8. Ibid. p. 87.

9. Ibid. p. 87.

10. J. Tyrrell. Report to MacKenzie on the Conrad Mines. Nov. 1906. Tyrrell Papers. University of Toronto. as quoted in Lundberg. Op. Cit. p. 90.

11. Tyrrell. Op. Cit. Perhaps, the differing viewpoints of Conrad and that of Tyrrell exemplify the differences in Canadian and American business mindsets-that of caution versus one hell bent for profit and tomorrow.

12. Lundberg, Op. Cit. p. 114.

13. Ibid. p. 117.

14. A.C Libby "Spokane Aerial Tramways Span Lofty Mountains" *Spokane Spokesman Review.* Nov. 13, 1927.

15. John Bufvers. "History of Mines and Prospects," Ketchikan District Prior to 1952. Alaska Division of Mines and Minerals. 1967.

16. Ibid. p. 6.

[17] Bufvers history is hopeful, yet factual: "burned down," "snow slide," "disappeared," "explosion" and "drowned" are a few descriptions used. Ibid. p. 19.

[18] As stated there were more than fifteen aerial trams on Prince of Wales Island. Other than the Riblet installations, the builders names are not known for the remaining trams. Ibid. p. 19.

[19] BC Johnson. "Copper Deposits of the LaTouche and Knight Island Districts." *USGS Bulletin 622.* 1918. p. 192.

[20] Ibid. p. 214.

[21] Ibid. p. 215.

[22] Thanks to Nancy at Prince William Sound Books, Valdez, for this information.

[24] See W. Maxwell. "The Anyox Concentrator of the Granby Company." *Engineering and Mining World.* April 1930.

Enter the Guggenheims

31. The Guggenheims

No history of mining in North America can be written without remarking on the Guggenheim family. No one family influenced mining and spearheaded the change from nineteenth century lone mining operator to that of large, industrialized twentieth century syndicate more than the Guggenheims. By the period of the First World War, the Guggenheims controlled around seventy five percent of the world's base metal output. What is more astonishing is the fact that the family only entered into to mining business by default, and at a late date too.

The Guggenheim clan were Swiss-German-Jewish immigrants who arrived in the New World in the middle of the nineteenth century. The family was originally from Guggenheimb in Bavaria, though the family moved to a more stable Switzerland due to the effects of the Thirty Years War. They were firmly established in the town of Lengnau by the end of the eighteenth century. At first prospering in the Swiss Cantons, the family succumbed to lower middle class status. Such was the state of affairs when Simon Guggenheim was born. He worked as a tinker and tried to support his family. This he did for several decades until 1847.[1]

On the eve of the year of revolutions, Simon Guggenheim opted for a more prosperous future and emigrated with his large family to America. They settled in Philadelphia and established a haberdashery. There they succeeded and did quite well. Much of the prosperity was due their expansion into stove polish sales. One of Simon's sons was Meyer. He was to become the patriarch of the dynasty. Simon died in 1869, and so ascended the 41 year old son, Meyer.

Mining was dropped into his lap by fate. A business associate could not honour a debt, and thus gave title to a thriving Colorado lead mine. It turned out that the mine contained much more valuable ore than was first suspected. This became the seed for a huge enterprise. With the rich ore as capital, he bought other mines in the area.

Meyer Guggenheim's family included seven grown sons when the Colorado mine came into his control. In the process, he and his family started to focus more on mining than millinery for the mines were producing some $17,000 a month.

> Meyer worked the mines himself from his office in Philadelphia, assisted by 130 miners and supervisory personnel in Leadville, and by 1887 they had produced 9 million ounces of silver and 86,000 tons of lead. Two miners on a single 12 hour shift could pull down enough ore from the stopes to pay all the mines'

expenses for one day. The miner's went on strike time and again, and time and again the strikes were broken, usually with armed state militia or hired thugs. Meyer never had much patience with striking workers. He himself had been a worker, the lowliest of workers, and he had never struck, he only worked harder and harder.[2]

In total the Leadville mines yielded $15 million to the Guggenheims.

To further increase profits and reduce costs, the Guggenheims went into the smelting business. By treating their own ore, they reduced their unit costs and thus made more money.

Another expansion the Guggenheims presciently resided over was Mexican mining. Daniel Guggenheim was sent to Mexico City and he extracted mineral and tax concessions from the Mexican President Porfirio Diaz. Essentially, the Guggenheims were given a free hand in Mexico. Thus they built smelters and began exploiting the rich silver, copper and lead mines of that troubled country.

However, there were obstacles in business, namely competition. And it came in the form of a rather formidable player, that of J. D. Rockefeller. Rockefeller had diversified his Standard Oil interests, had bought mineral rights and had started mining companies. The gold mines of Monte Cristo in Washington State were one of his projects. By both horizontally and vertically integrating the industries, Rockefeller desired to control the smelting business. He planned a large trust—the American Smelting and Refining Company—to effect this end.

The American Smelting and Refining Company (reduced later to ASARCO) first tried to drive the Guggenheims from business, by controlling supply and price. Rockefeller had done this in the oil industry. However, things did not work out as planned. Due to strikes and market fluctuations, ASARCO suffered on the stock market. In the process, the Guggenheims sent in agents and silently bought up majority shares of stock.

The end result was that Rockefeller lost controlling interest of the company he created to destroy the Guggenheims. Ironically, the Guggenheims triumphed.

> When the smoke cleared the American Smelting and Refining Company and the Guggenheims were one. Daniel Guggenheim was chairman of the board and president. Solomon was treasurer. Isaac, Murray, and Simon were members of the board...and the Guggenheims and their allies owned 51% of the stock.[3]

ASARCO became a jointly owned company between each competing sides. But with such financial backing, the world was these New York financier's oyster.

The Guggenheims devised an unbeatable formula for success:

1. Develop when the markets are in the trough of the business cycle.
2. Always use the cheap labour and raw materials of undeveloped countries to depress your own country's industries to force its wages and prices down until they are so cheap you can afford to buy them up.
3. Own everything: own the mine, smelter and the marketing.[4]

As one can see the Guggenheims used their size to bully and monopolize the metal industry.

Another reason for the family's success was the division of labour between Meyer's sons. Daniel Guggenheim was a visionary, and could spot opportunity. Isaac and Murray ran the dry goods business. Benjamin graduated from the Columbia School of Mines and was sent to oversee mining operations in Colorado. Solomon was sent to manage the Mexican operations. William, in turn, became the partner in the Colorado Smelters. Simon studied Spanish and was later sent to the South American interests. Control of the expanding empire of M. Guggenheim and Sons centred in New York. They eventually became one of the richest families in the US. With it came the trappings of wealth: Park Avenue homes, estates on Long Island, private yachts and railcars, and, eventually, philanthropy.

Meyer Guggenheim died in 1904. Before his passing he beseeched his family to stay together for strength. The Guggenheims were at the height of their power; with that power, they would control the copper market.

As stated, the Guggenheims started development of the massive Kennecott copper mine in Alaska. It was a gigantic project. J.P Morgan was brought in to help finance it, and then:

> **they bought Kennecott Mountain and several hundred thousand acres of adjoining territory in the Wrangell chain, bought two hundred miles of railroad right-of-way to the sea, bought seacoast land at Katalla Bay, Valdez and Cordova, bought the Northwestern Steamship Company...bought every Alaskan coal mine they could get their hands on, bought endless forests, and perceiving that ships could transport fish as well as copper, bought, as a sideline, Northwestern Fisheries.**[5]

The railway alone from Cordova to Kennecott cost $25 million (in 1910 dollars). The St. Elias terrain was unforgiving: some of the highest mountains in North America, the region also had the most snowfall, and little flat land. The Copper River and Northwestern Railroad was built [by Michael Heney the same contractor on the White Pass and Yukon Railway] over the river to find flat land and avoid glaciers. The railroad took four years to finish, and when completed ore was shipped to Tacoma for refining.

One gigantic enterprise was not enough for the Guggenheims. In 1911 they bought and began expanding the copper mines in Bingham Canyon (Utah) from underground to open pit operation. It would become the second largest mine in the world.[6]

Not included in these large projects were the many small holdings that the Guggenheims, and their companies furthered. ASARCO owned wholly or portions of the most productive mines in British Columbia. The Sullivan Mine was partially owned by ASARCO, and the Premier Mine in Stewart, at the time the leading gold producer, was completely owned by ASARCO. The mines at Keno, Yukon were owned by ASARCO as well. Many smaller mines in BC were Guggenheim interests too.

Soon their investments were expanded to the rich mines of South America. These included the Braden and Chuquicamata mines in Chile; which became other aces in their stacked deck.[7] But the windfall was around the corner. When the First World War erupted in August, 1914 the Guggenheims controlled the copper market. Copper was an essential material for bullet and shell casings, electric wire, and machine bearings. Lead was also needed for bullets and ships ballast. "In 1916 alone American Smelting and Refining sold $234 million worth of metal to Britain and France."[8] The First World War lasted for four and a half years. Correspondingly, the American support of the Allies was as much to do with business as it was with politics or submarines.

M. Guggenheim Sons bought interests in Angolan and Congo diamond mines. They also industrialized the Yukon River placer mines. The large gold dredges on that river were financed by the Guggenheims and they proceeded to upturn vast rivers of the Canadian Yukon in a successful search of that yellow metal. The rich tin mines in Malaya also drew the Guggenheims. They formed Pacific Tin, and installed river dredges to sift for that alluvial, stannous metal. The cartel also bought the Bolivian Carocales Tin operation in the high Andes.

The Guggenheim business philosophy included hiring capable men. It put leading mining engineers on its sizable payroll to find, develop and bring into production valuable ore bodies. They also were quick to see and install new technology that would, in the end, return more dividends. The Riblet Tramway Company was seen as a leader in its field and thus ASARCO and other Guggenheim mining companies hired them to install tramways at their projects. This is how Riblet was contracted for the Premier and other mines in the Canadian West.

Yet business success for the Guggenheims could not avert tragedy from visiting the family—Benjamin Guggenheim drowned in 1912 when the *RMS Titanic* sank. Then William Guggenheim challenged his brothers in court over money from the Chilean ventures and consequently aired their private dealings in public. Then in 1922 there was an inter-generational dispute within the family over the 51% interest of ASARCO shares. In effect, the elder family members wanted to cash out, while the younger ones did not. Eventually, a portion of the shares were sold, to pass on controlling interest, though the Guggenheims still owned blocks of shares in the company.

With such resources the family began to live out scenes from The Great Gatsby. By the Jazz Age, Meyer's sons were aging and the grandsons were taking over. Issac Guggenheim suffered a nervous breakdown, obviously from carrying the burden of being the eldest son, and others were passing on. However, the old guard had a few more tricks to play: that of leading edge philanthropy.

With such colossal wealth, the Guggenheims supported good works. It started small at first, with a wing to the Sinai Hospital, then it flourished with archaeology in Crete, and backing of Robert Goddard in his quest for reliable rockets. A large bequest was a "fund for the promotion of aeronautics" which established schools and projects in that young science. Other gifts were made to musical, educational, and public health institutions. A large gift was made to New York University as well. The Guggenheims also had brief ventures into the field of politics. Simon Guggenheim bought a seat in the Senate, when he was successfully elected to represent Colorado in 1907; while Harry and Robert Guggenheim were later to become Ambassadors for the United States.

Daniel established the Daniel and Florence Guggenheim Foundation to pursue the written arts. Many books and scholarly works have been supported with money from this foundation. Solomon Guggenheim capped the lot with the most visible and remembered endowment: that of the Guggenheim museum by Frank Lloyd Wright.

In the 1920's Solomon was coaxed by his new wife to purchase a sizable collection of Modern Art. As a result, he wound up with a warehouse stacked with Kandinskys, Legers, and Chagalls. To display the collection, the Solomon R. Guggenheim foundation was established. Its purpose was to build a museum to display the paintings. Solomon personally requested Frank Lloyd Wright for the project, and the plans for the Guggenheim Museum were approved in 1944. Unfortunately, the project was not to be so expeditiously completed. After many delays, controversy, arguments, the deaths of the patron and architect, and disputes with the New York City planning board, the museum finally opened in 1959.

Thus the legacy of the Guggenheims spans immigration, mercantilism, mining, monopoly, marine disaster, money, and modern art museums. It is quite a record.

32. Braden Copper Company, Chile

A mere 50 miles from Santiago, and its adjoining port of Valparaiso, is Chile's second largest copper deposit—Teniente.[9] In the nineteenth century, Chile had been a world leader in copper production, providing nearly one half of the world's supply in 1870's from deposits at Teniente. Copper was a steady cargo—together with wheat—for the windjammers on the European run. Thirty years later, the Chilean copper market share collapsed. The reason for the slide being due to North American production, a drop in the concentration of the Chile deposits, the use of much hand labour and low capitalization.[10] By the turn of the twentieth century the North American capitalists were looking for new prospects and places to invest their profits. With the low wage scales, minuscule tax rates and huge copper deposits, Chile became ripe for the picking. A man named William Braden—who had ASARCO connections—started the first North American venture in that dry, mountainous country. In 1904 he controlled the Teniente copper deposit. In course, the Guggenheims were brought in as partners. And later "the Guggenheims formed the Braden Copper Company in 1908 with the capitalization of $23 million."[11] While rich in content, mining was difficult as theTeniente was at 7000 feet elevation. Whole camps had to be built from scratch, as did railways, waterworks, and the social support systems.[12]

The Riblet Tramway Company erected a mining tramway at the Braden Mines about 1905. The tram was four miles in length and had a large, 100 ton per hour capacity. Heavy four wheel bucket carriers and double grips were used. Details on the tramway are few, but it is known that the cableway worked for a few years until the miners punched new tunnels within the hill—then the ore was carried within the mine haulage systems—and thus negated the need for the tramway.

The Braden Copper Mines was a large operation. The mines had 15 Shay locomotives moving the ore and railway cars around. This is in addition to an unlisted number of rod locomotives. Consequently, one can see the extent of the undertaking. The Braden Mines were profitable, however, and with such profits the Guggenheim's company Chile Exploration became interested in other rich Chilean copper mines. This is how the Guggenheims became involved with the large Chuquicamata deposit in Northern Chile.

To this day, Tienente is still a large producer. Canadian expertise still plays a large role in mining operations in that Republic. Many Canadian mining companies and suppliers have operations in Chile.

33. Stewart, BC.

Another major mining region in BC is the Stewart District. Many important mines have operated there, and continue to do so. Stewart is 150 miles north of Prince Rupert, at the head of the Portland Canal, a 70 mile long inlet that separates BC from the Alaska Panhandle.

Not only a political boundary, the Stewart Region begins the North country and has northern topography and bio-regions. The mountains dominate the region—huge peaks erupt from the sea and rise steeply to seven thousand feet. Those peaks are formed with stark, grey granite walls that are defined with sheer rockfaces, avalanche chutes and mountain alder. Massive glaciers and icefields carve the valleys and cascade in huge ice rivers down to their toes where they form glacial rivers. This is the beginning of the glacier belt that extends to Glacier Bay and the St. Elias Range. Everywhere is wet. Stewart sits on the confluence of the Pacific Storms and the Arctic weather. The mountains also form their own weather, and force the storms to drop their torrents of water. Stewart has a large rainfall of around 80 inches per year. The forests drip and the grey rock is made greyer with wetness; everywhere the water cascades and runs off the mounains, into creeks, streams and rivers. In the summer months the granite mountain walls shimmer in the sun from their water coating as water thunders off the wet peaks and melting icefields.

That water feeds a damp, primordial forest of cottonwood, hemlock and spruce. The dark grey is set off with a deep green. Moss and dense undergrowth such as devils club and false hellebore define the groundcover. Grey and green colour the place—the grey gravel and rounded river rock form the ground. in more exposed places, the grey glacial till is heaped up in an occidental Offa's dyke. That grey ground, more often than not, has grey overcast sky. Over this rock flows a persistent wind. As the valley is a conduit for air between the Coast and the Nass Valleys and, as the sea air and glacier air form constant convection currents, steady winds flow.

In the winter, life is even more exposed. The persistent rain falls as snow. Stewart has one of the continent's highest annual accumulations of snow. It is regularly recorded in the order of one hundred feet total annual snowfall. That snow feeds the glaciers. And for much of the winter snow sits on the ground. No, snow is the ground as the snow is piled meters deep. It has been known that the citizens must tunnel through the snow out of their homes. In the winter the grey and green subside into a monochromatic black and white vista. The snow being the white backdrop which turns the trees and rocks an illusionary black.

Miners had been exploring the region for a few decades: for instance, there was a gold rush in the Stikine in 1862. While there was another, later rush in the Omineca. Mineral wealth was suspected to be in the area, only the remoteness and ruggedness of the terrain hindered the avaricious designs of men. Stewart first came on the mining scene with the arrival of Klondike prospectors. Sixty-four Seattle prospectors were dropped off at the muddy estuarial head of the Bear River. Laden with promises of gold, they naively disembarked and tried their luck at placer mining.[13] They found traces of gold but no large riches so some made their way over the ice of the then large Bear Glacier and into the Nass Valley.

Three men stayed on and located the first silver claims—D.J Rainey, Pap Stewart and John Stark. "Rainey locating the Roosevelt group of claims at Bitter Creek, and, a year after, 'Pap' Stewart and Ward Brightwell locating the American Girl on American Creek."[14] Both creeks are only several miles up the Bear river from the tidewater setting of Stewart. John Stark, a settler from Saltspring Island, found other claims on Glacier Creek. The waterfront land was pre-empted in 1903 and by 1905 lots were sold. Other claims were staked on Glacier Creek: the Jumbo, Hallie, Apex, Evening Sun and Red Cliff.[15] A cable bridge was installed in 1907 to aid crossing the large Bear River.

The region was coming into being. Men and women ventured north from the now depressed Kootenay region seeking new territory. For a few years Stewart became the promising camp. A stamp mill and concentrator were installed at the Jumbo property on Glacier Creek. It also had a short "Bleichert" tram erected to feed the mill. While the Red Cliff Mine erected an 800 foot tram and concentrator at American Creek; the Brown Alaska property shipped ore in 1906.[16] In the process, settlers put up tents and log cabins. A townsite sprang up with houses and hotels on streets of gravel till.

Meanwhile the Little Joe and Lucky Seven Claims—owned by the Portland Canal and Development Company—installed a 2400 foot tramway and fifty ton mill in 1907. Stewart was just finding its feet by 1907 when the economic malaise hit. As a result, development stalled for three years. This was the first stop in the stop and go cycle in Stewart's history.

Despite the downturn, the Portland Canal Short Line Railway was incorporated and set about to build a line fourteen miles up to the American Creek Mines. The potential prosperity of the region attracted some big players. Donald Mann sent his agents to investigate, and

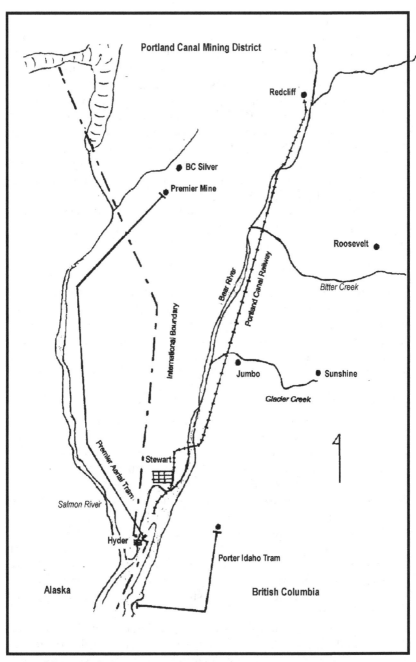

Map of Stewart, B.C.

he became a keen supporter of the region. In 1910 the railway was finished, which was subsequently bought by the MacKenzie-Mann railway syndicate. With and eye to extending rails over the mountains to the Nass Valley, this group even dreamed of stretching it as far as the Peace District financed on visions of timber, coal, wheat and minerals.[17] The railway hoped to develop into Canadian North Eastern Railway as it expanded in the promoters ever more fanciful minds. With such acceptance in the hallowed halls of promoter's money, Stewart expanded to include the famed King Edward hotel, a hospital, theatre and even a stock exchange. By 1910, a huge wharf was stretched on piles one mile out into the estuary (to locate deep water). From there the coastal steamers could unload their heavy industrial equipment. Ships from the Grand Trunk and Pacific Railway and Steamship line based in Prince Rupert called in.

To the southwest, at the Salmon River mouth, Stewart's siamese twin settlement in Alaskan Territory sprouted. A tent city there was legitimised and christened Portland City, and later Hyder after a Canadian geologist. The border divided Portland Canal, when it intersected the meridian and turned North just two miles before Stewart and the Bear Creek estuary. Of course the establishment of the Boundary in 1905 had no connection to the local geography. The meridian goes hither and yon, up mountains, and down valleys. Because of this, when the border was established it prevented easy, all Canadian connection to the upper Salmon River, Unuk and Stikine Rivers. It was a decision that would perpetually affect both the Stewart and Hyder populations. Hyder boomed in its own right with the establishment of the Riverside mine by the Salmon River some 5 miles north of Hyder. With this, Hyder expanded after 1917. A townsite was built on piles over the Salmon River estuary. Legend has it that Hyder was a wide-open place with much liquor drunk and carousing undertaken.

Stewart, on the other hand, had its boom in 1910 but by 1913 the financial markets were drying up. That year was a bad one financially, and the war the next year permanently changed the business landscape. The previous ever-expanding markets rationalized and it would have a large impact on BC. The Canadian North Eastern Railway idea was shelved. Even the half-finished Pacific Great Eastern stopped construction for its projected use (that of connection of the Grand Trunk and Pacific Railway from Prince George to Burrard Inlet) never materialized. After the war the Grand Trunk and Pacific Railway and the Canadian Northern Railway would be bankrupt and sent by a federally funded receiver into the conglomeration

An angle station on the Premier Tramway—Stewart, B.C.

called Canadian National Railways. Against such a background, activity in Stewart dropped off again. Men joined the ranks, mines shut down, and only a few prospectors remained.

At the end of the war prospectors were locating rich veins north of Stewart—on the Upper Salmon River, some 20 miles distant by road. The ore seam headed in a north south orientation, and while the earlier prospectors had located outcroppings in the Bear River valley, the seam reappeared to the north. These properties became the Big Missouri, BC Silver and the Premier claims. Pat Daly, an intrepid grubstaker, discovered the Premier Claim in 1916. By 1918 mining prospecting increased to a flurry and Stewart again became a hive of activity. It turned out that the Guggenheim clan had bought the Premier mine and began to develop it.

As the Premier Gold Mine was some twelve miles distant, north of Stewart on the Salmon River, a tramway was needed. Special concerns for this system included, capacity for the snow load, acute angles on the line for it to follow river systems, and the fact that the tramway crossed over the Canadian border into the US and then recrossed back into Canadian territory only at the end of its journey. For ten miles, the tram would be on US soil, and thus the ore and goods in transit were placed into bond for customs purposes.

Another problem included having to power the tramway from the minesite, as that was where the generators were. Because of this, the traction cable would have to be continuous for its entire $11^1/_2$ mile length. Also, the tram rose up a hogsback hill before its final descent to the lower terminal and wharf. This was not to be a gravity system. Preliminary surveys were conducted in 1920, while application for right-of-way use applied for with the Provincial and State governments.[18] In 1920, contracts were issued to clear a hundred foot wide swathe through the dense coastal forest. Massive trees were encountered, and their bulk and brush had to be moved from the line.

Looking south on the eleven-mile Premier Tramway in the 1920's. Snow was a problem.

Royal Riblet's 1921 Diary has an entry for Feb. 21. "Submitted bid on Premier Tram." A later note on March 1 states "Mr. Peters [of ASARCO] phoned me to come to the office and said we were 10% high." Then on March 3, the entry simply notes "Completed deal. Offered job on that basis of reduction."[19] And so the Riblet Tramway Company was contracted to build the longest and most complex tramway operating in North America at the time.[20]

A division of labour was decided by Premier Gold Mining Company, the Guggenheim business agency set up for the operation. The Riblet Tramway Company would do the design work and supply the hardware; the mining company would do the construction. This colossal structure would have 152 wooden towers, four angle stations and two terminals. The line would be 11.5 miles long and rise from sea level to 1300 feet at the minesite. Two fifty horsepower electric motors would be needed to power the single travelling cable at fifty feet a minute.[21] That cable was 5/8ths of an inch in diameter and twenty-three miles long, while the track cable was one inch diameter on the heavy side and 7/8 inch on the light.

The tramway gauge was eight feet, measured between track ropes, and ground clearance was a concern due to the excessive snowfall. The cables travelled some thirty feet off the ground, less when the snow lay on the ground. A road was built and the materials hauled in with truck and horse. During construction, the ground was blasted to bedrock, for suitable foundations for all structures; local timber was milled nearby and used for the towers, angle stations and terminals. Towers sat on wooden sills, without the aid of cement. When the towers were up, the cables were strung using a strawline, block purchase, and horse whim.

The one hundred twenty in number, eight hundred pound capacity ore buckets were of standard riveted sheet steel design and were spaced two minutes apart on the line. The grip had a lever activator which allowed the bucket to engage and disengage at the angle stations as the travelling cable rounded sheaves. Men assisted the transfer of the bucket through those systems. Two men were on duty at each of the terminals. Maintainance and tower men rounded out the crew; Pitt, the Premier Mine manager, elaborates that "The entire 3 shifts require 31 men."[22] He then goes on to discuss day to day problems.

> Being on 24-hour operation, and having such severe-weather and operating conditions to contend with, naturally wears out cables, bucket sheaves, tower sheaves, and the bucket tubs. Replacements are made as required and charged direct to the operation. Track cables average about four years' life, while traction ropes average about three years. The heavier ropes now in use will give longer life, however, Sections of track rope over the ridges where grades change abruptly fatigue more rapidly and require more attention. Linemen grease all tower bearings several times weekly. Buckets are oiled daily in the terminals, and the travelling cables is likewise oiled by oil dropping on it from a barrel equipped with a suitable dropping device. A special line oiler equiped with a pump actuated from the moving bucket sheave is periodically sent over the line to oil track cables.[23]

The maintenance regime was similar to that of other large tramways.

Consideration had to be given to the angle and tension stations. In order to direct the cables to the line of the meandering Salmon River, three large angle stations were built. One was of acute measure at 81 degrees; the others were at 153 and 133 degrees respectively.

Large, roofed wooden buildings housed the angle track, wire rope guides and sheaves. They were elevated off the ground at the running height of the cable, and allowed the buckets to pass through, detach from the traction cable, change direction and carry on after the building. There was an attendent to oversee the largely automatic operation—he had to ensure that the buckets grips properly detatched and reattached. Within the angle stations, the track cables were held taut first with loaded weight boxes. Three additional crib anchor stations were located every 2/3 of a mile, where one end of the track cable was firmly fixed. At the next station there would be a block purchase and geared winch which allowed for adjustment of the heavy cable. Load and thermal variances altered the deflection of that cable. The block purchase was found to be the best method of tensioning, and later fully replaced the cumbersome weight boxes.

The lower terminal was a two story large wharf structure, built on piles, with tram terminal above, with ore bin, and warehouse below. A conveyor belt allowed for easy transfer of the ore to waiting freighters.

The tram moved 400 tons of concentrates each 24 hour period when in operation. The Premier Tramway cost $360,189 when finished in late 1921. As stated it was the longest aerial tram in North America, only the La Dorada tram in Colombia was longer. (See Appendix VIII.) Also, for many years the Premier Mine was the largest gold mine in BC. A fire in 1928 tried to take that title when it reduced the cookhouse and support buildings to ashes. Only the quick work of the men prevented the fire from spreading to the mine, concentrator and tram buildings. The mine soon resumed operation.

To alleviate from the stress of mining and transport, entertainment could be found at Hyder. Men would ride the trambuckets for a quick ride 'out', (usually a forbidden activity) and would jump out into the snow before reaching the lower terminal and discovery by the tram operator. Some miners disappeared into the dance and gambling halls of Hyder for days in a Bacchanalian bender. To further these vices, an enterprising business-person set up a camp closer to the mine at Eleven Mile Camp on the Salmon River right over the international border. A cabin was built spanning the border with a half-inch gap for that imaginary line to pass through. During the 1920's, when the Volsted Act was in full force, liquor was prohibited in Alaskan territory, thus the need for two halves to the cabin.[24] A bar was set up in the Canadian corner, while a bordello was legally established on the Alaskan side. It was a full service premises and which also obliged the laws of both countries.

Premier Gold Mining ran for many years until 1953. It was a large camp with hundreds of men on the payroll. A vast townsite existed at the town of Premier to house the men. Supplies were shipped in from the wharf near Hyder on the tramway, in via shoeshoe shod horses and sleigh, or Crawford Transfer Company caterpillar tractors and road train. Even then men would be sometimes camp bound due to the snow. The Depression did not affect the community as much as other places, as people remained employed.

The Porter-Idaho Tram

In the 1920's Stewart gave life to another remarkable tramway, not as wellknown as the Premier operation; nonetheless, this installation was in many respects the most challenging of the two. "So successfully has the Premier tram met the requirements for transportation to and from the Premier [mine], and so satisfactory has been the operation from the cost standpoint" that Riblet Tramway Company was engaged to build another difficult tram in the Stewart District. Due south of Stewart, high on the Marmont Basin was the Idaho claim. Surrounded by glaciers, and situated just beyond 5000 feet the mine was, in 1927, pushing the limits of accessibility. Flush with success from the Premier operation, the Premier Gold Mining Company bought the claim. In turn, they set about developing the property.

At first pack trains ascended the mountains from the beach at and they allowed for a rudimentary measure of transportation. For full production a tramway would be required. From the beach at the estuary of the Marmont River, the tram would slash its way south through the forest. After two miles, the tram would change course with the North Fork of the river and ascend further into the mountains and glaciers. High cliffs hindered easy access and placing of structures. In the end, a large unsupported span of just under a mile was necessary.

The siting and construction of this tramway was difficult. Glaciers, high cliffs, avalanches and brutal weather limited structure locations and made special provisions necessary. Because of the topography, only a rudimentary pack trail was available; a fact that made the use of large, local timbers of wood in the trams problematic. Unlike the Premier tram, the Porter-Idaho tram was constructed of steel.

Due to the large unsupported span, the tram was broken into two separate trams—an upper and a lower—with an intermediate station at Green Point connecting the two portions. The lower portion also contained one angle station to correct for alignment: "From the beach to Angle 1 there are 33 towers, 2 tension stations, and 1 short rail section; from the angle to Green Point there are 2 towers and 1 tension; and beyond Green point there are 2 towers, 2 tensions, and 3 three-rail sections."[25] In addition to the tension stations for the heavy track cable, the rail sections were needed to account for the loads of the heavy spans.

The Lower Terminal was 130 feet long and 30 feet wide. It was a massive four story structure, with cable anchors, bullwheels, ore bins and adjoining wharf.[26] Bunkhouses, a diesel powerstation and support buildings were also situated there.

Lower station and warehouse of the Premier Tramway near Hyder, Alaska. Ships loaded here.

The tower steel was $3^1/2$ inch angle iron, with standardized designs and fabricated at the beach, then bolted together on location as needed. The towers sat on 15 inch diameter by 8 inch concrete feet which were cast on site (sand, cement and water needed to be hauled in).[27] Rocks would be then piled atop the discs to hold them down against the natural movement of the tower when the cables and buckets were in motion.

Byron Riblet's method of 'pioneer drive' was employed to sling the steel and supplies from the beach to the worksite. A small portable winch powered by a car motor would activate this construction tramway. It was planned to install the track cables up the towers from the beach to provide a method of transport for the higher towers and structures. This way the tramway would leapfrog forward using the tram to move the steel. Only a delay in the delivery of the heavy cables negated this plan and forced the crews to labouriously hand balm and pack the materials in. As a result, a road was cut for the first two miles to transfer materials, then the supplies were transfered to the pioneer drive for their movement to Green Point. Another smaller tram continued for a few hundred more feet, after which the materials were horse packed on go devil sleds, and packframes over the trail and glacier to the upper regions.

The heavy $1^3/8$ inch locked coil track cables were moved in on spools, afterwards they would be unwound using 1/2 inch straw lines to haul them over the ground to the towers. Once extended they would be raised up the tower with fall blocks, secured in the saddles and stretched to tension with the tension stations. The same method was used for the smaller 7/8 track cables. The large track cables were secured to large buried steel frames which acted as immovable deadman anchors.

Once the lower tram section was in operation, steel could be moved more easily to the upper section. And there was much material to move: there were two tension stations to build, and the large upper terminal. It was 105 feet long and of similar construction to the lower terminal, the main difference with the angle being of heavier steel to withstand the heavy snowload, and possible snowslides. It also provided support for the mine operation.

> **There was still lots of work on the rail sections and tensions on this upper end...[and] many tons of material had to be dragged to them. The raising of steel to a height of 90 feet out on a hillside with a slope of anything from 30 to 90 degrees, and with winds up to fifty miles an hour blowing at intervals, taxes the endurance of men and slows the job. The frontier produces men to meet such conditions, however, and eventually they accomplish these jobs. And so by meeting these and many other trying situations, the tramline, the powerline, and telephone line were all completed during the summer [of 1928] and are now in operation.[28]**

The tram was five miles long when completed, and owing to the construction difficulties cost $312,000, nearly as much as the Premier tramway. And, remarkably, no one was killed during the project.

Owing to the steepness of the second part of the tram, double carriers and grips were used on the buckets. Each bucket had an eight cubic foot capacity, and was equipped with automatic swivel grips.

Pitt then goes on to describe a journey in an ore bucket on the Porter-Idaho tram.

One thrills as he looks out at the rugged mountains on either side, iwth their dense covering of spruce and hemlock, and their jagged snow covered peaks reaching high up in the blue sky. Peculiar sensations are stirred when one looks over the side of a bucket, nearly a thousand feet down into a river, and when the glaciers, filled the deep canyons or creeping fan like down the mountain sides, come into view; then the real grandeur of the scene impresses itself. As one rides smoothly over the line and follows with the eye the old trail over which horses and men, sweating under their loads, have laboured many hours ....[29]

The Porter Idaho tramway was an impressive achievement for the Riblet and the Premier Mining Company and one that they were understandably proud of.

Maple Bay Mine

Thirty-Five miles south of Stewart, down the Portland Canal, was another mining camp. There, the Granby Mining and Smelting Company operated the copper mine at Maple Bay. It was worked from about 1910 and sent ore to the Granby complex at Anyox. Details are sketchy on this operation, although it is known Riblet built a two mile tramway to convey ore down the mountain to an island-like terminal and ore bin that sat on piles in the channel. From there the ore was moved by ship around to Observatory Inlet. Apparently, the Maple Bay mine worked for several years during the First World War and into the 1920's. Little is heard about it after this period.

An undisclosed Riblet lower station and wharf: probably at Prince of Wales Island, Alaska.

Notes —9

[1] John Davis. *The Guggenheims*. Morrow and Company. New York. 1978. p.40.

[2] Ibid. p. 60.

[3] Ibid. p. 75.

[4] Ibid. p. 110.

[5] Ibid. p 103.

[6] Canada's largest mine is the Highland Valley copper mine near Ashcroft, BC. It is a vast open pit, and was in business from 1955 until 1999. It was another ASARCO interest too. (See: LeBourdais. *Metals and Men*. p. 85.)

[7] Determining who owns what mine is difficult. While the business pages say a mine was owned by "The Blue Sky Mining Co." that company could be really owned by the Guggenheims. Then there is the position of minority blocks of shares, and the Family simply investing in other mines. In fact, the Guggenheims became circumspect about the degree of their holdings after they were pilloried inthe press because of the overt purchases of Alaska. For example, a 1942 book on geography in Latin America is very guarded about the end owner of industrial operations there. Its descriptions of owners are so fleeting to simply say "and the copper business [is controlled] almost exclusively in New York." Preston James. *Latin America*. Odyssey Press: New York, 1942. p.261. The wealth and degree of power of the Morgan-Guggenheim syndicates became an embarrassment in the thinking circles and we saw how it ultimately became one source of trouble for Chile in 1971 and the tragic *coup d'etat* of 1973.

[8] Davis. Op. Cit. p. 121.

[9] Valparaiso has fifteen famous funiculars to the town. They work to this day.

[10] Thomas O'Brien. "Rich beyond the Dreams of Avarice-the Guggenheims in Chile." *Business History Review*. Spring 1989. V.63. p. 8.

[11] Ibid. p. 9.

[12] See Harry Guggenheim "Building Mining Cities in South America." *Engineering and Mining Journal*. July 1920. Vol. 110. No. 5. pp. 204-206.

[13] Ozzie Hutchings. *Stewart*. Solitaire Press. Cobble Hill, BC. 1976. p. 6.

[14] Ibid. p. 6.

[15] Ibid. p. 6.

[16] See www.stewartbc.ca

17. Dreams of crossing the Omineca were not limited to just one rich industrialist. Charles Bedaux, the French Taylorist time and motion expert and Nazi, sought ways to spend his millions in the Depression. He made his fortune improving factory production in Europe and America. He then proceeded to kit out an expedition to cross roadless Northern BC from Edmonton to the sea in Citroen half-tracks. A nouveau French Aristocrat, he travelled with champagne, crystal, caviar, and courtesans. His entourage provided a boom and a much needed distraction to the Depression ravaged Peace district. In the end the forests proved too much for him and the Citroens; they were pushed over a ledge and the expedition was forgotten. He also went on another safari to the Congo in proto-RVs, purpose built by the International Company. The Congo basin was too much for that outing, as well. He reappeared again in the Edward and Mrs. Simpson abdication crisis as Bedaux provided the French Chateau for their lonely wedding. Bedaux cast his lot with the Nazis and Vichy France, and, apparently, he was grooming Edward Windsor for re-installation on the British throne after its conquest by Germany. Edward was a closet Nazi, too. When Germany occupied France, Edward headed for expensive exile in Portugal where he was even involved in an Abwehr scheme to bring him to Germany. After this he was sent to his Elba in the West Indies by the British government. Bedaux died in a US jail where he was put as punishment for his Nazi collaboration.

18. D. Pitt. Aerial Tramway Construction and Operation. *Transactions of the Canadian Institute of Mining and Metallurgy.* 1930. p. 1139.

19. Royal Riblet Diary. Cheney Cowles Museum.

20. The Riblet Tram at Encampment, Wyoming tram was 16 miles in length, though it had ceased operation in 1908; the tram at Pecos, New Mexico was 11.7 miles long but was not built until 1927; while the Peruvian tram was 20 miles long and finished in 1928. All were built by Riblet. In 1953, the BRECO Company of London would assist in building a 50 mile long system in Gabon, Africa.

21. Pitt. Op. Cit. p. 1142.

22. Ibid. p. 1130.

23. Ibid. p. 1150.

24. As told by John Ledo, Stewart pioneer.

25. Pitt Op. Cit. p. 1160.

26. Ibid. p. 1163.

27. Ibid. p. 1166.

28. Ibid. p. 1172.

29. Ibid. p. 1172.

Washington State

34. Chewelah, Washington

In the unsettled Northeastern corner of Washington State is the Colville River. Acting as a natural north-south conduit, the mining men travelled the Colville Valley often back and forth from Spokane to the Canadian Kootenay Mining regions. It was only logical that adjacent Washington State mineral resources were developed. Populations were drawn to the area as it is surrounded by rich and productive agricultural lands. In time, the town of Chewelah flourished about the turn of the twentieth century.

High quality magnesite deposits were located at Finch Quarry. It is a mineral used to make refractory bricks that were necessary to line the Bessemer steel furnaces to prevent their walls from melting in the intense heat. Previously, the material was obtained from Austria in Central Europe. With the US entry as a combatant into war, new sources were necessary as Austria was now at war with the US. As a result, the Northwest Magnesite Company was organized during the First World War to develop the Chewelah area deposits. B.Thane and R. Talbot, of Spokane and San Francisco, were the principals involved.[1] The mining company was to become a solid customer for Riblet Tramway Company, ordering six separate tramways over the years to assist their quarrying operation.

Duelling tramways; parallel tramways at Metalline Falls, Washington.

The first Finch Tramway was 5 miles long and hauled 60 tons of material per hour from the pit to the plant. It was built in late 1916 and had a long life lasting until 1968. The tramway was such a success that Northwest Magnesite had Riblet build a second tram immediately parallel to the first. The second tram was built in 1934, "the right of way [cleared] of heavy timber at very heavy cost because of the terrain."[2] Trams were also built to the Redrock and Keystone claims which were adjacent to the Finch operation. A thousand foot long waste delivery tram completed the installations all at the quarrying operation to the southwest of Chewelah. The ore was quarried and hauled to the plant in town where it was further processed. The remains of the works can still be seen to the south of town.

Other local magnesite producers hired Riblet as well to erect trams for them. Riblet built a tram just to the south at Valley, Washington.

Ferry and Stevens County, Washington

The region of Washington State—adjoining Grand Forks and Rossland on the Kettle and Columbia Rivers—had several large mines. Indeed, the towns of Republic and Northport were established for and prospered by mining and smelting. Two mines of interest were the Electric Point mine near Northport and the Laurier Mine at Laurier.

The Electric Point mine was a lead mine northeast of Northport at the hamlet called Leadpoint. C. Johnson discovered the mine and its free lead outcropping in 1912 and he, in turn, sold the mine for $22,000.[3] With the demand for base metals in the First World War, the Electric Point Mine became a sizeable camp and operation. The mine returned some $3 million dollars in dividends in 1917.[4] First, the ore was shipped by horse and wagon and later a cableway was installed. To capitalize on this mine a Riblet tramway was built to move the ore to the camp of New Boundary. Another Riblet tramway in the region was built at the Laurier mine in Laurier, Washington. It was seven thousand feet long and built about 1902.[5] While the First Thought gold mine in nearby Orient also produced ore valued in the hundreds of thousands of dollars. To facilitate such output it installed a 75 ton per day capacity tramway. The Canadian butcher Pat Burns, later owner of the Blakeburn colliery, was its principal owner. Sadly, there is little information available on any of these undertakings. What is known is that northeastern Washington State was the home field for the Riblet Company, and there were many examples of their fine products there.

35. Metaline Falls, Washington

In the extreme northeast corner of Washington State is the attractive valley and village of Metaline Falls. A resource community with lumber companies and a Cominco lead mine that operate today in the region; previously limestone production featured in the town's history. The town itself was built to support the local mines.

Nearby limestone deposits facilitated the production of cement, activity that was started by the Inland Portland Cement Company in 1910. The Inland Company quickly sold its operation to the Lehigh Cement Company, an eastern concern and they operated the plant for eighty years. Under Lehigh's management large cement silos were built and an aerial tram constructed to move the limestone to the plant. Their Riblet tramway was four miles long and was capable of moving 60 tons of rock per hour. There is no listing on when the tramway was built.

The plant was busy for many years supplying cement for the nearby dams and the large Grand Coulee dam. The Washington and Idaho Railway terminated at Metaline Falls and connected the town with Spokane and other points to the south. Consequently, it was able to move the bags of cement to market. A hydroelectric powerhouse was built on Sullivan Creek to supply the needs of the cement plant and town. Nearby, other hydro plants operate on the picturesque Pend Oreille River. The town's namesake waterfall is over a striking box canyon on that river.

The French consortium of Lafarge Cement bought out Lehigh Cement Company's Metaline Falls operation in 1990 and it subsequently phased out operations there. Thereafter, Hollywood and Kevin Costner came to town in order to make a mixed bag of a movie about mailmen. Today only the large silos remain of the cement operation while the town restores its old houses and buildings.

36. Ione, Washington

Logging was important too in the rugged northeastern corner of Washington State. The region contained many resources—sited there were the minerals that lay underground, and large timber stands as well. And, it was not trucks that moved the timber, but something else.

Those timber stands brought Eastern capitalists who sought other sources of raw timber after the wood in the east had been cut. Washington contained wonderful stands of fir, pine and cedar and those stands of white pine brought in competing companies to Pend

Orielle County. Two companies—the Diamond Match Company and the Panhandle Lumber Company—greedily eyed the virgin timber.

The Panhandle Lumber Company had first moved to the area in 1907 and had built sawmills in the region. This included a large mill at Ione and land holdings on the lower LeClerc River from where they had built a connecting lumber flume to move the timber from their holdings on the LeClerc River to their mill. The newcomer Diamond Match Company had also quietly bought up pine forest in the LeClerc Valley too and intended to bring it to market. This company used the fine, straight-grained pine to cut into blanks for wooden matchsticks. The blocks would be prepared into flats and then cut into sticks. Finally, the sticks would be dipped into sulphur and phosphorous to make the completed match. These last steps were usually done at urban match factories who employed young women to do the hazardous and tedious chemical preparation.

In Washington State, the Panhandle Lumber Company was determined to block the upstart company's progress by controlling the water rights for flumes and land access for railway lines. This rivalry began in the teen years, though only came to head in the early twenties. Another problem was that the timber lay on the eastern bank of the Pend Oreille River. A small wooden ferry carried traffic across the riverbut it would be unable to move industrial volumes of wood. At the time the Panhandle Lumber Company seemed assured that its rival was under control.

This was not the case in 1920, the Diamond Match Company had cut 5 million feet of trees and had built a small mill in the forest to mill it into rough boards.[6] In 1921 the Diamond Match Company hired the Riblet Tramway Company to build a $3^1/2$ mile aerial timber tramway to move the rough pine cants from the upper LeClerc River over two ridges and across the Pend Orielle River. The lower terminal would land on the western bank of the river about halfway between Ione and Ruby, and adjacent to the Milwaukee and St Paul Railroad tracks. With $1^1/8$ track cables and 3/4 inch traction cables, the aerial tram was in operation by July of 1921. With the tramway in operation, the business bottleneck at the LeClerc River mouth was solved.

While much of the technology was identical to the mining trams, the Ione lumber carrier used pairs of carriages which were chained to the 400 board feet packages of pine. In this way the carriers acted as slings for the lumber loads. Each carrier used the standard two truck carriage wheels and traction rope grip. The loads were sent on their way 800 feet apart; it is also claimed that the tram could move 10 million board feet an hour. The trams used a later issue wedge driven swivel grip.

Timber trams were not as common as the mining tramways, though there were a few around. One was built by the Trenton Iron Works, at Spanish Peak, California in 1918 and ran for about ten years. Another timber tram put together from used Trenton and Riblet mining tramway parts in Arizona in 1925. The mining tramways themselves often moved large amounts of timber for pit props and rough lumber, so timber carriers were not that unique.

At the same time the Panhandle Lumber company built a 3'–6" gauge logging railway and bought some new Shay engines to move the timber. The company then proceeded to build rail lines up the LeClerc River into the timber. Logging operations—like mining operations—usually only lasted a few years until the resources gave out. This was true at Ione too. With two lumber companies cutting in the valley, there was not much longevity for these operations. The Diamond Match Company shut its operation in the summer of 1927 and the aerial tramway was sold to an unknown buyer.[9]

37. Sumas and Concrete, Washington

While European settlement came early to beautiful Whatcom and Skagit Counties in Washington State (people settled there just prior to the Civil War), the rugged mountains and wide rivers hindered wholesale development. Bridging the Skagit River had been a problem and it caused a lag in development. An early bridge at Mount Vernon had been washed out in 1894. There were other rivers and sloughs, too, which further inhibited transport and trade.

Nonetheless, logging, mining and fishing did prevail. By the turn of the twentieth century, industrialization was well advanced. The prospectors had scoured the North Cascades looking for mineralization. There were a few mines, the largest was at Monte Cristo. Another, the Bornite Mine at Darrington, installed and used a half-mile tram in 1903. It was owned by the Tacoma Company, the proprietors of Vananda Mines at Texada Island. Also, huge deposits of very high-grade limestone were discovered in Skagit County. Likewise, it is about this time that engineers stopped building with stone and wood and switched materials to use steel and concrete. In a word, exploitation of those limestone deposits resulted.

By 1908 the capitalists were circling, namely in the form of Seattle Electric Light and Power Company. They had hydroelectric designs on the Baker and Skagit Rivers, and to build these dreams, cement would be required. Thus the town of Concrete was established and limestone quarrying began.

The limestone was situated high on a mountain, thus transport was needed. A heavy capacity industrial back and forth tramway was built to move the blocks of limestone to the crushing plant and silos in the town.[8] Unlike the small bucket carriers of the metal mines, this operation had large prism buckets, and it was capable of carrying 250 tons of limestone per hour. So far, other details on this operation are few. The terminals were built of concrete, as were the silos. Afterwards, the rock was then powdered and sintered into quicklime, from which cement was made.

The cement was used at the Ross and Grand Coulee Dams. Also, a young contractor named Henry Kaiser (later of Hoover Dam, Liberty Ship hull and healthcare fame) attempted to improve the muddy roads of Skagit County by laying a cement surface on a portion of them in 1910.[9]

Grotto, Cement and Civil Suits

On the topic of cement, the Olympic Portland Cement Company had a similar operation near Sumas, Washington. It had a mine on a hillside and two linked to and fro trams down to a silo. Leschen built the tram in the 1920's. The limestone was then hauled away by trains via a connecting Milwaukee, St. Paul and Pacific Railroad spur from the appropriately named Limestone Junction. While further south in Washington State, where the Skykomish Valley runs through the mountains, Highway 2 crests near the whistle stop of Grotto. It is a hamlet of infinitismal size; a cement works was established there in the 1920's. Riblet built a 150 ton capacity to and fro, calcium carbonate conveyor there.[10]

Interestingly, cement later became an irritation to Royal Riblet at his picturesque estate. Along with the fine house and grounds, he also built a small, tourist tramway over the Spokane River to access his house. However, his site by the river attracted other people too. A cement plant set up shop upwind of his house. By the late 1930's

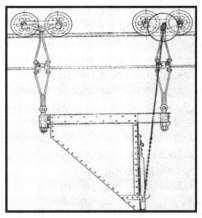

A Riblet skip bucket similar to the one used to haul limestone at Grotto, Washinton. The chain is a trip.

and 1940's the cement plant dust was constant—the dust would settle on his house and tramway. In the end, Royal Riblet filed a lawsuit against the company for negatively affecting his house and life.[11] The case meandered through the courts and was unresolved at the time of his death.

38. The Snake River Trams

The Columbia River Basin is a large geologic feature spanning Oregon, Washington and Idaho. It is an immense basalt lava flow that makes table lands and buttes in the region. Subsequent volcanic eruptions have covered these flows with layers of ash, which make for rich farm lands. Wind blown loess also coat the hills. Across this strata, the watercourses have fought their tortured way, and in the eons of elapsed time, erosion has left spectacular cliffs and escarpments.

With the arrival of the European, patterns of transport and trade were organized along riverways either with boats or railways following the river courses. The European also set about exploiting the region's resources. This basin was turned into an immense granary—the volcanic soil yielding bounteous wheat crops. One problem was getting the grains to market.

The wide rivers and steep basalt cliffs created some difficult obstacles. To this day, the access roads down to the Snake and Clearwater Rivers are steep and winding. Early on, the farmers slowly
and labouriously shifted their wheat sacks with wagon teams down the torturous grades. Yet Victorian ingenuity set to work—funicular inclines were built around the region—a famous one was built at Mayview, Washington. There was even an attempt to chute the grain down the cliffs in pipes until it was discovered that the resulting friction from the descent cooked the seeds in their husks. The farmers thus turned to the nascent aerial tram. A Huson model was put to work at Wawatai, Washington, and a Bleichert model hauled grain to the elevator at Kendrick, Idaho. There were others too around the Northwest wheat regions namely Kelly's Chute, Judkin's Chute and Beale's place.

Riblet built a two and a quarter mile long tramway at Lenore, Idaho some two score miles east of the Snake River. It was built in 1903 and moved two hundred wheat sacks per hour on its modified ore buckets. It operated until 1937 when the receiving grain elevator caught fire. The flames caused further mishap for as the tram station burned, their cables dropped onto the Camas Prairie Railroad. In due course, a train fouled the cables demolishing the engine's chimney and whistle. The tram was not rebuilt after the fire.

With this background, the farmers at Peck, Idaho faced a similar freight problem. Peck was a small town some thirty miles east of Lewiston on the Clearwater River. The regions claim to fame was the major placer gold strike at Pierce in about 1863. This fostered the Idaho gold rush, whereby, some American miners from the Cariboo gold fields, rushed to Idaho, and many others came (including Chinese miners) in their eternal quest for gold. Transport was via the Clearwater River. After the gold rush Peck became a mixed industry town with lumbering and wheat farming up on the plateau. The Northern Pacific Railway stretched its Camas Prairie branch up the Clearwater River, but on the north bank—a feature which was be a reoccurring problem as Peck was on the south bank.

There was a small cable ferry across the river, but it was slow and could not handle heavy traffic. One farmer did not mind the steep hills that he had to handle hisloads on, it was the inclines off the ferry that caused him headaches. Something had to be done. Magill and Gurnsey, the local owners of the grain elevator, elected to raise a tram to carry wheat across the 800 foot river from one silo to another. Riblet installed the tram and it had a 200 sack per hour capacity.

In the end the citizens of Peck were so aggrieved by the transport situation that they even passed a bond issue to raise money to built a bridge. It was finished in 1918 and served them well until it washed away in the floods of 1948.

Byron Riblet built another tram nearby at Moscow, Idaho. Details on this installation are few other than it was raised in 1921 for the Nesbit and Wells Company. Owning to the agricultural region it too was probably used to move wheat.

Notes—10

1. It appears that Northwest Magnesite was involved in a lengthy court case over unauthorized mining on State land, The court case records much detail, and thus acts as an important historical source. (See *State v. Northwest Magnesite*. 1947.)

2. Ibid. p. 69.

3. Northport Over Forty Club. *Northport Pioneers. Statesmen Examiner*. Colville.1981.

4. Ibid.

5. Lake, Ruth. *Kettle River Country*. Statesman-Examiner. Colville. 1970. p. 172.

6. See Bill Laux. "Diamond and Panhandle-a Timber Rivalry on the Pend Oreille." *Timber Times.* Spring, 1995.

7. The Diamond Match Company was headed by a man named Stettinius, who, amongst other things, headed a First World War Industrial Production Board. His son would become US Secretary of State during the Second World War.

8. This was a heavy duty continuous tram. It has not been determined if the tram was of either Leschen or US Steel origin.The author suspects it was the latter; either way it was not built by Riblet.

9. Henry J. Kaiser, like Byron Riblet, started in Spokane. He later moved to BC and eventually California where he flourished with the Hoover Dam and other New Deal Projects.

10. Phil Woodhouse notes a large tram at the King Mine, Grotto; it may have been moved from the cement operation. (See Woodhouse *et al. Discovering Washington's Historic Mines.* Vol. 1. p. 151.)

11. For details on Royal Riblet's pollution court case see William Barr. "Man against the Corporation." *Northwesterner Magazine 31.* 1987.

The U.S.West

39. Oatman, Arizona

Fifty miles east, as the crow flies, from the Colorado River and the California border is the northwestern region of Arizona—an area whose claim to fame was a gold strike at Goldroad in 1902. Nearby, was another instant town—that of Oatman. Both of these communities were boomtowns in a gold rush of the early twentieth century.

A Riblet tramway over a town; this link hauled gold ore without incident overhead at Oatman, Arizona.

Gold was discovered at Oatman below a geologic monument called the Elephant's Tooth in 1900. After a series of owners, the mine (first called the Blue Ridge and later the Tom Reed) became productive and gold valued at $133,000 was removed. As a result, more scientific, and deep-pocketed operators moved to the district. George Long examined the geology and suspected the ore vein to thrust northwards from the Blue Ridge and he staked the ground. In the process, valuable ore was found and a new rush was on.

The United Eastern Mining Company set up shop and proceeded to develop the property with a cyanide plant and tramway. Thus, about 1916, the Riblet Tramway Company was contracted to build a short tram one mile long connecting the mill down the sage hills to the town. In fact, the tram passed right over the town on high pylons en route to the ore bins, and it did so without incident. The tram had wooden towers and had a 50 tons per hour capacity. Today, the derelict tram towers still exist, albeit in a desiccated state, snaking their way from the town to the mine.

As with most mining camps the boom soon passed and Oatman slipped briefly into obscurity. Later, in the 1920's the famed highway Route 66 passed through the town, as it does to this day, resulting in continued life for the gold camp. Oatman survives as a way stop on that outmoded but historic road, as a curiosity for vacationers and filmmakers.

40. Yosemite, California

Immediately adjacent to the western entrance of Yosemite Park is the small hamlet of El Portal. In addition to the tourism that the National Park created, the local economy also consisted of timbering and mining.[1] Barium mining had started in the region in 1880's, and the completion of the railway to Yosemite furthered that industrial expansion. After several operators mined around El Portal in the early years, finally the Yosemite Barium Company bought the property. They began to develop it by the Jazz Age in order to sell the metal as an oil-drill slurry.

The barite ore deposits were plentiful in the hills, though they ranged in elevation from 2000 feet to 4000 feet altitude.[2] In order to move the ore from the diggings down to the mill and railway, a Leschen Aerial tram 3400 feet in length was installed.[3] It carried the ore from the mine to the north side of the Merced River. There it was processed in the mill—it screened, classified, and pulped the ore into barite concentrate. The operation worked for many years until the 1950's when it was closed.

41. Jarbridge, Nevada

The State of Nevada has a lengthy and interesting mining history. The silver mines of Virginia City are probably the most famous and remarked upon mines of Nevada. Mining, of course, went through various waves of development—as mining does to this day—with the early underground mines, to the later copper and gold mines of Ely, Beatty, Rhyolite, and Pioche. Today, mining continues in the north of the state and, rancorously, at Yucca Mountain near Las Vegas.

With this tradition, prospectors searched the arid valleys and mountains of that scorched land for other mineral wealth. In northeast Nevada, north of Wells and nearly abutting the Idaho Border, are the rugged Jarbridge Mountains. (Sometimes the name is in the form of Jarbidge.) A neglected and unknown region, it found purpose first with the large sheep boom of the Great Basin area during the early years of the twentieth century.[4] That changed with the discovery of gold by D. Bourne and J. Escobar in the Jarbridge Canyon in 1909. As Jarbridge was a remote gulch, access was difficult and land ownership was problematic because the claims were situated on Humboldt National Forest land.[5]

These difficulties were resolved with a road, and small townsite while other problems of a new mining camp arose. Properties changed hands into the larger players, some mines were overvalued and oreless, and new mineral discoveries found. Business confidence ebbed and flowed and with it. So did the extent of Jarbridge's architecture and arch-criminal class. A town and industry arose while robberies and holdups also became more brazen. (Jarbridge has the notoriety of being the site of the last stagecoach robbery in the West in 1916.)

Jarbridge boomed between 1916 and 1920. The Long Hike operation alone produced a million and a quarter dollars worth of ore. With it being such a valuable property, or course, ASARCO was involved. Also, it was to the Long Hike mine that Riblet supplied an aerial tramway. Similarly, Riblet built a second tramway at the Elkoro mine, which operated through the decade until the depression. Unfortunately, there are few details on the Jarbridge mining operation.

Jarbridge slowly withered after the depression due to declining industry and distance from cities. In the last few decades, it has become a hunting holiday hamlet and summer cabin region.

42. Tererro, New Mexico

The Rocky Mountains and the Pecos River form the geography for central New Mexico. Their rolling peaks, canyons and the pine forest covering made the country rugged. Roads and trails meander through the area accessing the hills. Mining had been part of the history of New Mexico. With sites such as Soccoro, and adjoining areas forming a mineral rich territory. The nearby states of Arizona and Chihuahua had huge mining industries.

In late 1925 the American Metal Company of New York purchased the Pecos Mine and intended to develop it from a base at Glorietta. Transport was a problem. The engineers debated the various available methods of transport. The winding, dusty service road, while able for the movement of light supplies, was not up to the sustained loads that the mine demanded. This was pre-pavement era with slight two-wheel drive trucks which stopped with mechanical brakes. In a word trucks were still in their infancy. It was not until after the Second World War that rugged, reliable truck technology appeared. When it did, the era of railways and aerial trams was finished.

A word must be said about road building ability. At the turn of the century, in the era of hand labour and horse scrapers, roads were primitive. Systematic methods of drainage, bridging, or water sealing were not developed. Railroads sustained the heavy freight. In the Jazz

Age, the first bulldozers, backhoes and mobile cement plants had appeared and thus a series of hard topped, Interstate roads appeared it the US.[6] The US embarked on a series of roadbuilding which was only curtailed by the collapse of public purses in the Depression.[7]

The American Metal Company then considered building a narrow gauge railway to service the Tererro mine but problems of assembling the required land along the line and the associated costs negated the railway idea. The company then elected to use the cheaper and more reliable method of aerial tramway. A request for contracts bids was tendered. Royal Riblet travelled to New Mexico and did a sight inspection for the tram, when bidding on the contract against US Steel, Roebling and A. Leschen and Sons.

In turn, the Riblet Tramway Company was issued the contract. Thus an aerial tram was constructed by our frontier freight funambulists "from the mine on the South Bank of Willow Creek, 12 mines to Alamitos Canyon, the site of the complex that would comprise the 600-ton mill and power plant."[8]

The terms of the contract were as follows:

Specifications:

General: An aerial tramway to be built for the American Metal Company of New Mexico. Length of line from Terminal end to Terminal end, 62,000 feet. The Tramway is to be built in two sections with a double control station at the angle point between the upper and lower sections. The difference in elevation between the loading terminal and the control station is 525 feet, the loading station being the lower. The difference in elevation between the control station and the discharge terminal is 1000 feet, the discharge station being the lower elevation. The capacity of the tram is to be 62 1/2 tons per hour, the capacity of each bucket 1250 pounds, and the speed at which the buckets travel will be 500 feet per minute.

Materials and Equipment:

Track Cable: The track cables on the loaded side will be 1 3/8 inches diameter plow steel, smooth coil, track strand breaking strength 105 tons. The track cables for the empty or unloaded side will be 1 inch diameter plow steel, smooth coil track strand, breaking strength 50 tons.

> Traction Rope: the Traction rope for the upper or mine section to be 3/4 inch diameter 6 x 7 Lang Lay plow steel, having a breaking strength of 46,000 pounds. the traction rope for the lower or mill section to be 3/4 inch diameter 6 x 7 Lang Lay crucible steel, with a break strain of 37,200 pounds. Anchors and anchor ropes to be of sufficient size and quality to exceed the strength of the track ropes. Cable couplings to be of nickel steel, thimble and wedge type.[9]

The tramway was finished in late 1926 and in regular operation by 1927. The tram was 62,000 feet long and hauled 600 tons of ore per hour. A control station was situated roughly in the middle to manage the cable tension and speed of the carriers. It was staffed by a crew of ten people: four in the mine loading station, and three each at the control and lower terminal. There were also two maintenance men who travelled over the line tightening bolts, and oiling the sheaves and cables.[8] At the upper station the first man would inspect the incoming, empty buckets oil the trucks on the cable hanger, and move it to the loading station. The loader would charge the bucket at the loading bin and push it to the launcher. "Every 36 seconds a light would flash and gong would sound to signal the launcher to send another-bucket on its way."[10] Launching the 1600 pound parcels of ore and steel was a strenuous job and thus the men rotated through the task.

At the control station the buckets were detached from the track cable, sent through the station and relaunched on another section of cable track. The control station was to break up the 12 mile tram into 2 workable sections. As with the Upper Station, launchers were required for both the heavy and light sides.

Once launched the grips would grab the traction cable and carry the bucket forth.

> A lever arm with a cast iron 32-pound truncated cone weight at the outer end operated the cable gripping system. It was mounted between the two wheels and when rotated upward, caused the cable grip to release; when rotated downward, it activated the grip. The iron weight served two purposes; 1) it rolled along a horizontal spiral cam thereby raising or lowering the arm, and 2) its weight provided the force to securely grip the cable. Its 32 pound weight was multiplied to a force of over 3000 pounds per square inch at the cable grip.[11]

The tram mechanics had a difficult job. They had to ride in the buckets, dismount from those buckets and oil the traction rope sheaves, look for broken strands of cable, and tighten bolts on the steel towers.

A portable telephone was supplied and reports phoned in via the line on the tram towers. Periodically, the mechanic would miss grabbing the tower as he leapt from the moving bucket. Generally mishaps were few, as the mechanics became proficient in their job. There were other mishaps as tram travellers infrequently were left dangling in the buckets whilst the tram shut down for the night.

Major trouble came in the form of a derailment; a light bucket left the track rope, stressed the haulage rope and separated it. Apparently, high winds had carried the bucket into the tower. That was not the full extent of the damage—the problem bucket caught tower 69 and twisted it from alignment. The resultant whip lash of the parting cable made it spiral away for thousands of yards. "The broken ends were separated by about one mile, and 40 buckets were suddenly damaged as they plowed into each other—some damaged beyond repair."[12] But that was not the end of the crisis as the quickly released cable snaked and catapulted its way toward the control and lower station—in the same manner that a broken tug of war rope launches pullers end for end—four men were injured from the mishap, and one eventually succumbed and died.

With the injured men packed off to the hospital, the rest set about repairing the damaged tower. A new cable was ordered from the Leschen Company by telegraph and it arrived four days later. The cable was hauled to the site with reluctant mules and installed. Thereafter, the tram was shut down in periods of strong winds.

A word should be said about track cables at Terrero. They were changed about every two years, and the cable joints greatly assisted this on the heavy cable. Attempts were made to rotate the cable to reduce the wear on any one spot, though it was found to be only slightly effective. The temperature gradient from winter to summer severely affected the cables with thermal expansion. In the process, the track cables slackened which caused the buckets to race toward the mill at the lower portion of the tram. "Mr. Riblet personally visited the site to try and determine the cause."[13] He changed the adjustment mechanism on the track rope tensioning system so it was better able to accommodate the weather.[14]

The tram was very successful and only needed a few modifications for locale conditions. According to Anderson the tram worked for over ten years. We are fortunate to be left with an accurate record of this Riblet tram in the New Mexican Rocky Mountains.

43. American Fork, Utah

High in the mountains of central Utah, was the Live Yankee Mine. It sat above and looked down on the American Fork Canyon which was just west of Alta. These mountains were also near the town of Pleasant Grove. Started in 1870, the Live Yankee was one of the region's largest producers of gold. Wagons and sleds were first used to haul the sacked gold ore from the adits at alpine elevation down to a siding of the Union Pacific Railroad. Of course, the mine was relieved of its high grade ore and then left fallow. Owing to the winding and steep roads, better methods had to be found to ensure production.

American Smelting and Refining Company bought the mine in 1930 and immediately began to modernize it. The Riblet Tramway Company was hired to build the four and a half mile tram from "the Live Yankee Mine to what is now the parking lot at Tibble Fork Lake."[15] There were many benefits to the all steel, heavy tram. It could operate in all weathers and was also very durable in the harsh mountain conditions. And

> the tram eliminated ten miles of wagon or truck hauling down steep, curving upper canyon roads. Four large, especially built dump trucks were purchased to haul from the tram to Pleasant Grove.[16]

The tramway had thirty towers along with four anchor stations. "They were located on: Silver Creek, Porcupine Ridge, Major Evans Ridge and Mary Ellen Ridge."[17] The tram was finished at the end of 1931.

To carry the ore thirty buckets of thousand pound capacity were employed. The mine and tramway had long lifespans—both operated for twenty-two years until 1952. In 1932, three men were riding in a bucket when the carrier derailed and the men were thrown to the ground. Although bruised and injured, the men survived after treatment.

The Forest Service removed the majority of the tram in a clean up effort, though a few of the steel towers of the tram still stand on site along with a few collapsed mine buildings.

The mine notwithstanding, the neighbouring town of Alta entered tramway history. To offset the effects of the depression, Alta reconfigured a Trenton Aerial tramway into a bicable ski chairlift. It used tramway parts from the Michigan Utah mine.[18] Parts may have been borrowed from the American Fork operation, too. As a result, Alta sprang to life as a ski resort—a role that it flourishes at to this day.

Notes—11

[1] Indisputably, mining featured highly in California's history with the placer mining around Sacramento. However, mining continued into the industrial era with operations in the Grass Valley, Shasta, Bodie and in Death Valley. Naturally, there were many aerial tramways in California and it is indeed a topic that requires investigation. Some tramways have been documented while others have been forgotten. One long tramway that may have a Riblet connection was the Cerro Gordo tram (see footnote below.)

[2] Young, George. "Mining and Milling Barite-Operation of the National Pigments Company in Mariposa County." *Engineering and Mining Journal*. July 24, 1930.

[3] For information about the Leschen named tram at El Portal see Ibid. Whereas, the historian Bill Laux worked at this operation in the 1950's and he claims the Riblet Company constructed it. This could be another instance of joint participation by the two builders. Letter to Author from Bill Laux. October 2001.

On the leeside of Mount Whitney, in the dry Owens Valley, there were many mines and aerial tramways. Of interest, was the large installation at Cerro Gordo locaed a few miles south of Keeler, California. It was built to transport ore from the lead and zinc mines in the Inyo Mountains and it worked off and on from 1908 into the 1930's. The tramway itself survived until the 1950's when it was removed and is still remembered today as just one more curiosity of this unique region of Eastern California. In the Owens Valley Museum (Laws Depot) there is a large, blue enamel sign emblazoned with the words "Leschen Tramway." It came from the Cerro Gordo link. (See Robert Trennert. *Riding the High Wire—Aerial Mine Tramways in the West*. p. 82.)

[4] Basque shepards settled in this area of Nevada and Idaho in the latter half of the nineteenth century. Basque culture is famous in Idaho to this day. Of course, Basques were only one part of the Celtic migration to the new world-travellers that included Scots, Cornish, and Irish in their quest for a better life.

[5] Gibson.Jarbridge, Nevada.OnlineArticle www.hometown.aol.com/Gibson0817/jarbridge.

[6] While the American States embarked on their first interstates, Canada lagged behind. The roads of Canada were largely unpaved before the Second World War. When the first Trans-Continental Road was finished in 1926, it could hardly be said to be a road. It was an unpaved track. There were so few roads, and because of the terrain many were not connected—so much so that portions of the province drove different sides of the road as was the local custom. In the 1950's many highways were still rough and unpaved. The Trans Canada highway was not completed and hard topped until 1962. The current state of the Kicking Horse Pass highway, Hope-Princeton and the Fraser Canyon are indicative of the slow action on high way development.

[7] The headlong rush to the future is a common theme in the American Empire. We see it over and over whether it be canals, railroads, telegraphs, navies, cars, roads,

carfins, shopping malls, H-bombs, internet, telephone companies, or deregulated electricity distribution companies. It seems that there is not much thought is given to the longterm future.

Collapsed bridges were a common theme. The US Army led an expedition to travel from Chicago to San Francisco just after the First World War in the era of dirt tracks. It took the army months to get across its own country. Their heavy trucks and loads collapsed several rotten bridges. This lead to the "Good Roads" groups, the Pershing Road map of interconnected state roads and eventually the linked highways in the 1920's.

[8] Stanley Hordes. "Land Use Investigation, Forest Highway 17." Environmental Improvement Division, State of New Mexico; Santa Fe. 1990.

[9] The rest of the Riblet-Pecos contract specifications was as follows;

Carriers: Four hundred buckets of ten cubit foot od 1250 pound capacity. Carriers to be eqquiped with two wheel, car type, self oiling steel trucks, sheaves, or wheels to be fitted with removable treads. Each carrier to be fitted with gravity type self-adjusting grip.

Supports: All-steel towers and tensions. To be built of angles and channels, bolted assembly with lock washers where needed. All structures receive one heavy coat of red paint. All structures to be fully equipped with all necessary machinery, i.e. steel rocker saddles, cast steel idler sheaves with self oiler bearings on unit bases, double rope guides and rail sections over tension stations.

Loading and Discharge Terminals: To be designed and constructed to fit storage bins, and conform to plant layout at mine and mill, such features to be determined by design and construction of mill and mine plant.

Control Station: To be located at angle in line at approximately four miles form mine terminal.To consist of two ten foot diameter steel grip wheels equipped with steel automatic grips, main shafts with three vertical bearings and a heavy duty thrust bearing, two seven foot diameter nine inch face cast steel bevel gears with cast steel pinions, two pinion shafts and bearings, traction rope sheaves, shafts and bearings, terminal saddles, approach saddles, track rails, rail chairs, rail tongues, automatic bucket attacher and detacher and bucket guides. One hundred horse power back geared electric motor with a ninety two inch by inch face pulley on pinion shaft and a six ply twenty inch rubber drive belt, also one controller. One forty horsepower back geared electric motor with controller, also eighteen inch six ply rubber belt with pinion shaft pulley. All the above listed, together with all equipment and appliance necessary for the operation and control of the tramway, to be installed. All to be housed in a steel frame building or shed having sheet iron roof and walls and wooden floor. All machinery set on concrete foundations.

Cable Oiler: One force feed cable oiler for oiling both track ropes.Telephone Line: One telephone system of number ten galvanized wire, metallic circuit, supported on steel poles to extend from the mine terminal to mill terminal, together with 3 standard mine instruments and 3 field instruments.

Structure Foundations: All foundations for towers, tension stations and other structures to be of concrete of approved quality, placed at sufficient depth to insure against frost action. All excavations to be backfilled.

Safety Control: A solenoid brake to be placed on the main pinion shafts at the control station. The operation of this brake to be such that if the electric power should go off the brake is automatically applied and the tram stopped, or the speed of the tramway varies below the normal speed any predetermined amount the brakes are automatically applied and the tramway stopped. Speed indicators and traction rope mileage indicators to be installed at the control station, and a bucket counter and spacer to be installed at the loading terminal.

There were also time limits, completion penalties and bonuses in the contract. (See E. C. Anderson. The Aerial Tramway: As used in the transportation of Ore at the Pecos Mines of the American Metal Company of New Mexico, Tererro, New Mexico. Unpublished Thesis. 1938. p. 6.)

[10] Leon McDuff. *Tererro*. New Mexico, 1993. p. 42.

[11] Ibid. p. 42.

[12] Ibid. p. 43.

[13] Ibid. p. 44.

[14] With the longer trams of the 1920's, cable expansion (due to load or heat variances) became more pronounced. The Premier tram in Stewart had a cable purchase system to tighten the track ropes, though the daily weather in Stewart is similar to being underground in a mine- "cold, damp, and with low ceiling." New Mexico's sun would be much stronger. Ibid. p. 45.

[15] See: www.plgrove.org/historical/claydump.htm

[16] Ibid. p. 1.

[17] See www.americanforkcanyon.com

[18] This was a peculiar hybrid lift. One that used mining parts and telephone poles to make a ski lift. It worked for several years from 1939 until the location closed in 1941 for the duration of the war. Nonetheless, there is a direct causal link from the mining trams to ski chairlifts.

South America

44. Bolivia

Bolivia is integrally associated with mining. One silver mine at Cerro Rico underwrote the Spanish Empire in the sixteenth century. By the nineteenth century the important mines were the nitrate deposits of the Arica desert on the Pacific Coast. These nitrates were used for fertilizer and high explosive production for naval guns. So valuable was the nitrate that Chile had a war with Bolivia (War of the Pacific, 1879), which resulted in Bolivia losing its coastline, nitrates and seaports, a fact which its citizens have never forgotten. In compensation, Chile arranged for a railway to be built to Bolivia from the now Chilean port of Arica. [1]

By this time Bolivia's rich tin mines had been discovered. So extensive were the deposits that they have affected tin production in the world, and tin mining in turn had a large impact on that country's history. The person most responsible for the Bolivia's tin production was Simon Patino. He was the driving force behind the discovery, development and consolidation of the industry (after 1903). There were other tin barons too, though Patino is the best remembered. In the process he became fabulously wealthy and lived like a European aristocrat.

He had big mines and smelters at Uncia, Catavi, Oruro, Cochibamba and Llallagua. In fact, ore was moved at Llallagua by three Bleichert of Leipzig aerial trams from the mine to the works in the steep, dry country. Railways later replaced the trams in 1926 in a drive for more production.[2] In the process that mill's output was increased to 800 tons of ore a day. The English tin cartel, of which Patino was a part, succeeded with the large First World War demand for metal. Later, tin consumption soared with the demand for automobile tinplate.

As a result, the Guggenheims expanded into Bolivia in the 1920's. They also bought mines in Northern and Southern Peru (q.v.). This Andean expansion provided the setting of a most amazing Riblet aerial tram. Near Eucalyptus, Bolivia, the Guggenheims acquired the Caracoles mining property which was, spectacularly, at 14,000 feet.

Byron Riblet wrote about the project.

> A site was selected and material ordered for the proposed tramway system. About the time the tramway machinery, cable, and construction steel arrived the price of tin jumped from 32 cents to 60 cents and the mine owners were anxious to take advantage of this rise in price as soon as possible.
>
> The owners wished the tramway to be complete within 100 days if possible. It was $6^1/2$ miles long, varying in elevation from

9000 feet at the concentrator to 16,000 feet where it passed over the summit of the Andes, and then down to 14,000 feet, the most rugged country over which our company ever built a tramway.3

The 'pioneer drive' system of rigging a temporary tramway and leapfrogging the equipment forward was used. The tram was nosed forward over the distance. Another problem, beside altitude and altitude sickness, was the steep nature of ground. Detours and diversions around gaping chasms were necessary. Mules were used as transport around the rugged peaks. The Argentine mine itself was on the rim of old crater which was filled with water. In the end the tramway project was finished in 1923—and on the hundredth day of construction. Three other smaller tramways were also erected to haul ore from the Carmen Rosa, San Enrique and San Elena mines. The throughput capacity of the trams was fifteen tons per hour.

An ambulance unit was built to carry patients out of the Andean mines. A gurney was placed inside a sheet metal carrier, which was, in turn, suspended on two tramway brackets. Hauling out the injured and sick by tramway was much quicker than a rough, eight-hour, pack trail trip.

The ultimate fate of the Guggenheim Bolivian mines is not known. What is understood is that the expanded Patino facilities of the 1920's enlarged tin output near to the point of overcapacity. Compounding matters was the Guggenheim's entry in the market and its complete collapse due to metal prices in the Depression. All severely affected the Bolivian tin industry.[4] Afterwards, instability became the rule as metal prices fluctuated due to depression, market supply, war, cartels, labour strife, revolution, and embargoes. All have made the Andean tin industry volatile for the past seven decades. Nonetheless, the mining industry for many years was the backbone of that Republic's economy.

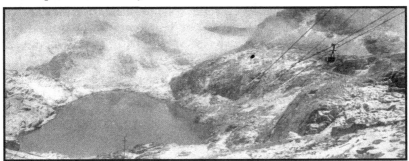

A Riblet tramway at work at a 14,000 foot tin mine—Caracoles, Bolivia.

45. Peru

In the mid-twenties the American Mining and Smelting Company's Northern Peru Mining division bought rich copper deposits near Samne in Northern Peru. To effectively produce at this mine an aerial tramway was needed to move the ore from the mine to concentrator.[5] A railway then hauled the concentrates eighty miles to the port of Chimbote. This mining development consisted of three properties—one copper, one coal and one gold and silver.

Byron Riblet goes on to explain the tramway:

> The country was rough. There was a difference in elevation between the loading terminal and the discharge terminal of over 5000 ft., so it was necessary to install and intermediate control station. Two angle stations were also required as it was impossible to install a tramway in a straight line between the terminals. From this first mine it was necessary to build a tramway 20 mi. long to the smelter site. Another tramway 2 mi. long was required to transport the ore from a copper mine to the smelter. A continuation of the tramway was required to reach a coal deposit about 5 mi. still farther up the mountain. The system, when completed, was 33 mi long and starting at 4000 ft, and reached an elevation of 14,000 ft. at the highest point, a difference in elevation of 10,000 ft.[6]

With such an altitude to consider, and the effects of altitude sickness, the construction of this continuously operated tram posed more than a few problems.

Riblet deftly solved the construction difficulties by building standardized tower sections in steel fabricated on site with local labour. The smelting company was building the smelter at the same time thus Riblet 'piggybacked' construction logistics on their freight waybill for a clever solution. Riblet devised "a temporary fabricating plant was erected, consisting of a hand operated shear and punch, gasoline operated hack saws, and ordinary gasoline drill press."[7] Jigs were used to allow for repeatability and minimizing the amount of measuring for the inexperienced fitters made up from the local labour force.

To haul the tower sections and steel to site, first wire then 1/2 inch cable and finally, one inch track cable was spooled out into a temporary construction cableway to the first tension station. There were 4 terminals, 3 transfer freight, 4 control stations, 2 angle stations, 5 rail stations, 40 tension stations and 308 towers."[8] A gas engine capstan provided the power for slinging the steel construction materials and cables to the next tower. These winches themselves were portable and travelled over the cables to where they were needed. The towers sat on concrete pads for support. These feet were "20 inches in diameter and 6 inches thick" and were precast to obviate the problem of hauling concrete mix and water to the arid sites.[9] Holes were dug and the concrete disks stacked to make a level and secure foundation.

With the foundations in place and a construction cableway overhead, the large track cables were then hauled out, anchored and used to haul material as well. In course, the towers were built under them and so the cables were raised in process. Tower measurements were taken during the construction, and those dimensions relayed by telephone back to the fabricating shop. In this way progress was fast and efficient: "The [construction] superintendent had a record in fabricating and erecting, completing nine towers, averaging 25 feet in height, in one day."[10] The Peru tramway was in operation in 1927 and upgraded over the following years.

The tramcars travelled at 500 feet per minute, which gave the tram a capacity of 12 tons per hour. The tram system started at the Callacuyan Coal Mines and descended to the Quiruvilca Mines, and thence to the mill at Shorey. There the ore was reduced into copper ingots for haulage the remainder of the 25 mile aerial tramway trip.

Riblet reaches the heights in the Peruvian Andes with interlinked trams between mines at Samne, in northwest Peru.

In the wild and rugged country of South America, an aerial tramway was the only feasible method of transport. Also, the scale and location of Byron Riblet's tramway for the Northern Peru Mining company in the high Andes literally marks the pinnacle of his success and it is easy to see why.

Notes—12

1. The nitrate deposits were also valuable for the European-nations during the First World War that the British Royal Navy and Germany Imperial Navy fought off Coronel, Chile. The Imperial German Pacific fleet was simultaneously attempting to return to Germany and help guard the German nitrate shipping from British blockade. Both beligerents were vastly interested in securing the deposits and bringing loads of nitrates back to Europe where it was converted into war materiel as TNT. In fact, early in the war the Central Powers were acutely short of nitrates. Only the timely development of the Haber-Bosch chemical process (whereby nitrogen was taken from the air and converted to nitrates) was the German war machine able to continue for three more bloody years. The German island colony of Nauru in the Pacific Ocean was another large source of nitrate. The island was one large nitrate deposit and was in course removed and sold. A Henderson cableway system assisted in the mining process. The island became an Australian protectorate after the First World War, and is now independent and living off the diminishing nitrate money.

2. R. Beard. "Property and Operations of Patino Mines and Enterprises at Llallagua, Bolivia." *Engineering and Mining World..* October 1930. pp. 545-550.

3. Byron Riblet. "Aerial Tramway Construction in the Andes" *Mining and Metallurgy.* July 1937. pp. 328-330.

4. Another lengthy tramway was built in Bolivia in 1936; furnished by the German firm of Pohlig, it was nine miles long and descended from a Mt. Auchanquilcho . (See Schneigert. *Aerial Tramways and Funicular Railways.* p.13.)

5. Again the true owner of the property is left open to doubt and not mentioned. Byron Riblet's article obliquely states that "The president of a large smelting and refining company once telegraphed me that they had acquired extensive mining interests in Northern Peru." Again, we see the effect of the Guggenheim's ukase against publicity. No date was given in the article for this tramway—it was about 1927. (See Byron Riblet's rare first-hand article "Aerial Tramway Construction in the Andes" *Mining and Metallurgy Journal* . July 1937. pp. 328-30.)

6. Ibid. p. 328.

7. Ibid. p. 328.

8. Wylie Graham. "World's Longest Bi-Cable Ropeway." *Engineering and Mining Journal.* July 1934. p. 287.

9. Byron Riblet is amazing as he is constructing large structures in the elevated and remote Andes and doing it in an age when industry had only just taken off its training wheels. In modern times, helicopters sling people and materials with ease. Riblet. Op. Cit. p. 329.

10. Ibid. p. 329.

The Depression

46. The Depression

The Great Depression marks a watershed for the Riblet Tramway Company and for Byron Riblet. After hugh success in the 1920's with the trams for ASARCO in Stewart and South America, the tramway industry shrank considerably. In the fall of 1929 the overinflated stock market collapsed. The large speculative bubble instantly contracted and people found their stocks worthless. Banks in the US went under in the thousands. Factory orders fell, and thus plants closed laying off millions. Life on easy credit from the jazz age suddenly became a brutal trial for survival. Markets were powerless to correct, and there was little Federal assistance to help redirect the process; monetary policy was a thing of the future. Great Britain, the previous bulwark of the money markets, was bankrupt. Its government was incapable and pursued a cautious 'safety first' policy.

Business in the mining industry evaporated. Overcapacity, falling prices, and few metal orders from closed factories all affected the resource industry. Miners were laid off because there was no demand for metals in workshops; thus mine exploration stopped (it required business confidence), and mine equipment suppliers closed. The tram builders Broderick and Bascom shut their doors while the Riblet Tramway Company very nearly followed them.

Like so many other self made men, Byron Riblet was stunned by the effects of the Depression. He was a man of retirement age and obviously reeling from the fact that the market did not want to buy his high quality engineering products. Like many men of his era, he did not understand that corrosive markets destroyed capable men. As a result he sought comfort in a bottle while the company drifted.

Apparently as an outlet for frustration Byron Riblet became slightly cranky. He used his motor car to block railroad crossings and delay trains, just as the trains had delayed his car; he drove at the speed limit in congested areas and impeded the flow of traffic. As a result, Byron was arrested for being a nuisance. Also, intriguingly, he took the time to telephone Franklin Roosevelt to try to give him advice. He also attempted to call the German Chancellor Adolf Hitler for the same purpose, although both times Riblet did not contact the heads of state.

Fortunately, Byron Riblet had engaged a draughtsmen named Carl Hansen. He was very capable and worked for the Riblet Tramway Company from 1919. (Hansen did leave from time to time to study engineering at the University of Washington and to work at the mines in Chile.) In the depths of the depression few orders for tramways were coming in. Byron asked Carl Hansen to 'mind the store' while he went on a sales trip. Strangely, Riblet's trip lasted almost eight months. Carl Hansen then essentially became manager of the Riblet Tramway Company. It was in debt and in a very shaky financial position. A few orders came in, notably the Windpass Gold mine and the Mt. Hood tramway but the company was in perilous state during the depression.

Both Byron Riblet and Carl Hansen put their personal money into the company to pay the monthly bills.[1] The difference of opinion that caused the rift between the two Riblet brothers stemmed from money during the depression. It seems Royal Riblet oversaw the sale of some spare tramway parts in 1933 and did not credit the company. As a result, Byron Riblet cut off his brother, terminated the thirty year partnership, and never spoke to him again.

Compounding matters was Byron Riblet's drinking. He would disappear for weeks into an alcoholic torpor. During the Christmas holiday of 1933 more tragedy visited Byron Riblet: his house on the Little Spokane River caught fire and was demolished. It is said that someone was careless with wrapping paper in a fireplace. All the same, the insurance money provided a welcome financial infusion. Riblet also sold some land from this large section of land on that river to further consolidate his holdings.

Under Carl Hansen's able control, the Riblet Tramway Company survived the Great Depression. More orders were came in, and the tidal wave of war work began in 1940. Hansen also oversaw the transition of the company from a mining supplier, to that of a general tramway builder, notably ski lifts. These lifts would go on to give a new market share for the company. In 1948 Byron Riblet sold his shares in the company (he held the majority of them) to Carl Hansen. The company from then on had no connection with the Riblet family. It must be noted that another young engineer joined the company in 1948—Tony Sowder. He was an Army Air Force veteran who would later rise to become a respected president of the Riblet Tramway Company. Carl Hansen stayed on as Vice-President.

Fortunately, the Riblet Tramway Company survived the Great Depression. It did so on the strength of new people and by seeking new markets. In a way, the old tramway company, and its products, were on the way out and would be replaced with a new line. Also, it marked the exit of Byron Riblet and his extensive legacy. The company would continue, but without its founder. An era had passed though the process would not be fully complete until 1950.

47. Windpass Mining, B.C.

Sixty miles north of Kamloops, on the Thompson River, is the hamlet of Little Fort. It is a waystop on the Yellowhead Highway with a few buildings and gas station, though the meager population does not indicate a paltry panorama. It is quite a salubrious spot. Five miles east of Little Fort on Mount Baldy was the Windpass Mine. That claim being discovered in 1916 by O. Johnson, T. Campbell and O. Hargen.[2]

In due course the Windpass Gold Mining company was established by Fernie businessmen in 1923. More ore veins were discovered and staked, though the property went through various owners without much development. After the market crash in 1929, the US dollar was detached from the Gold Standard. Later the dollar was devalued, when gold was placed at $35, and so the stable metal was much sought after by hungry miners. "Windpass reopened the mine in 1933."[3] Machinery was brought in to work the mine.

"An aerial tramline four kilometers in length was installed between the Windpass working and the north end of Dunn Lake, where a 50 ton per day mill was built."[4] The Riblet Tramway Company built the cableway and it provided much needed work for Riblet in the lean 1930's. The Windpass tram's capacity was ten tons per hour and the mine worked for the next five years.

The mine produced just over a million grams of gold with smaller amounts of silver. It also yielded about forty tons of copper. Various leasees optioned the property after the war and sampled the old workings and tailings piles. Texaco Resources and Kerr-Addison mines have held interests in the property in the last few decades and have conducted some diamond drilling. On the whole, though the property remains inactive.

48. Hedley, B.C.

In the stunning Similkameen Valley lay large deposits of mineral wealth. While that valley had long served as a conduit for miners travelling from the coast to other 'showings' such as Rock Creek, and Slocan, the development of the deposits was quite late in the realm of Canadian history.

Due to the Oregon Question, that is the question of ownership of the Oregon Territory, and the protracted withdrawal of Britain and eventual finagling of that territory into the American Empire, the Hudson's Bay Company were forced to reorganize their operations in

'New Caledonia.'[5] The HBC's loss of Fort Vancouver and Astoria, led to their formation of Fort Victoria in 1843. With that loss came the elimination of the easier trade trails in the now American territory of Washington. New Canadian trails had to be found.

Alexander Anderson then explored the rivers and passes in 1846 to discover alternate routes from Fort Langley to the prosperous fur depot at Fort Kamloops. He discovered the route of the present Hope Princeton Highway.[7] Edgar Dewdney was contracted to build a pack trail to the Gold Fields in 1860. As a result, travellers used the Similkameen Valley in the ebb and flow of trade.

Two generations later, the prospectors started to examine the region and not just pass through. James Riordan and C. Allison staked the earliest claims at Hedley in the 1890's. Others came in and staked nearby. Yet, despite the attentions, one sliver of ground remained unclaimed. A sharp eyed Duncan Woods, noticed the omission and secured it for himself. The missing claim—the Mascot—was found to hold a motherload of gold.

The mines on the mountain behind Hedley were developed from 1902, with adit, concentrator, and tramways. The Montana mining magnate Marcus Daly was behind the first impetus for development. Now, Hedley is unique in that it had several periods of development, much like at Britannia Beach. And like at Britannia Beach there were two portions to the operation. The concentrator and townsite was at the bottom of the valley, and the mining operation and support, located atop a mountain. At Hedley, the little camp of Nickel Plate was several thousand feet up the vertiginous slope. Connecting the two sites involved some engineering.

The first tramway was a famously long and steep incline. It had a surface skip pulled on railway tracks laid up the mountain side.

> This mine tramway, at that time the longest in the world, presented some interesting problems. Since the ore cars would have to be lowered on a continuous cable, it would mean the manufacture of a continuous cable 2 miles in length. No manufacturer was able to guarantee, let alone deliver, such a cable, and the solution was to break the tramway into two sections, each one half mile in length. Thus the mine cars could be lowered from the mine mouth to the half-way point, there transferred to the second tramway and lowered for the final journey down to the mill.[8]

The tramway fed into the ore hopper at the top of the concentrator. Also, the cable travelled over snubbing or brake drums to control the descent of the heavy cars. There were no brakes on the cars.

Hedley worked off and on for many years, stopping during the First World War and Depression. The tiny Mascot mine alone took out some $13 million in gold by 1955.

During the Depression, the mines were reorganized and equipment upgraded. "The resulting Mascot mine was constructed in 1936 with a skipline, to the mill via 20 Mile Creek."[9] The skipline was a single, multi-ton reversible tramway that was built by Riblet. The skipline tram ran from a new adit in the face of the cliff over a gully and out to the Nickel Plate camp. There, the ore was then reloaded onto the funicular for the thousand metre drop to the concentrator. It was quite an operation.

Mining ceased in 1955, and much of the mining equipment and builders were scrapped or sold. The funicular machinery was purchased for an incline near Port McNeill. The former Heritage Minister Bill Barlee instigated some conservation work to the wooden buildings around Hedley in the 1990's, and the Nickel Plate mine was reactivated atop the mountain, under a new company banner. Its operations are behind on Nickel Plate mountain, and accessed via a road. The old, wood mine buildings and townsite have passed due to fires and vandals, though they remained for many years standing on the hillside as a sentinel to the ingenuity, and steadfastness of the pioneers.

Not far from Hedley is the ski resort of Apex Mountain; many people hardly give a thought to the miners, engineering, or cablesystems that worked at Hedley as they whizz by for a day's skiing. They should, considering the connection of mining to skiing, a connection we shall see at another ski area.

49. Mt. Hood, Oregon

In the Northwest, the closeness of mountains, and the interaction of the people with the mountains in their logging and mining jobs, created a culture where outdoor activities became part of the lifestyle. This fact, in addition to the rise of mountaineering clubs and suppliers like Recreation Equipment Incorporated, furthered an alpine orientation. This period also saw the building of trails, huts, the creation of parks and the notation of first ascents of the various peaks. With the 1920's bringing in the first "Interstate" era in the US—one where cars are widely used on networks of interlinking paved state highways—and where cars were used to transport people for recreation; people had immediate access to the mountains. Earlier on, the railroads attempted to capitalize on mountain tourism with the building of large hotels such as Glacier Lodge and the Banff Springs Hotel in the Rocky Mountains. Yet these facilities were geared to the prosperous few. In effect, cars ended the railway's monopoly on mobility with their ownership of rights of way, ability to set ticket prices, expensive hotel rooms and make 'infrequent' schedules that catered to the those not on the one day weekend that the working classes laboured at until the Depression.

Another link in this chain was the work programs of the Great Depression: the Works Progress Administration and the Civilian Conservation Corps. These provided public money to build projects *pro bono publico*. In due course bridges, dams, schools, park facilities, the first skilifts, and post offices were built.[10] In the Northwest, paved roads were built to the bases of Mt. Hood, Mt. Rainier and Mt. Baker. Large, elegant lodges were also constructed. With cars and publicly built roads to the mountains, mass alpine recreation for all classes was possible.

Skiing had become a popular pastime by this time, and while people's skiing abilities and equipment were limited, having fun in the snow was the idea. Indeed, about this time, Hjalmar Hvam, of Portland, invented a safety binding—thereby raising the amount of skiing fun by lowering the number of broken bones. Numerous rope tows had been built to transport people at several areas, the first in Mont Tremblant, Quebec. A beginning skier's experience with coordinating skis, balance and a spiralling rope was usually disastrous. The diabolical and legendary Austin rope tow at Mt. Baker stood as testament to this: not only very fast, it ascended an impossibly steep slope. Hence, out of this situation we saw the birth of the ski chairlift.

The first ski chairlift in the world was at Sun Valley, Idaho; a winter playground underwritten by the Union Pacific Railroad to increase railroad passenger traffic. Their engineer, James Cullen, used wooden poles to carry the chairlift cables and it was in operation in 1937.[11] American Steel and Wire (a later reincarnation of Trenton Iron Works) was a consultant on this project. Similarly, the depressed mining community of Alta, Utah saw an opportunity to use the snow and old mining equipment to create business traffic. They reconfigured a Trenton Iron Works mining bicable tramway as a chair lift in 1939.[12] It was a small start, but the rest is history. These lifts and the Riblet passenger tramway at Mt. Hood some 60 miles from Portland, heralded a new age and an entirely new industry—that of commercial winter sports.

The Mount Hood tramway was a diversion for the main Riblet business. The Forest Service issued a Request for Proposals in 1937 to build a purpose-built lift at Timberline Lodge at Mt Hood. Carl Hansen, then senior engineer at Riblet Tramway Company, implored Byron Riblet to submit a tender. Riblet was astonishingly uninterested, as his mind was still on building larger and larger mining tramways. He also did not want to cooperate with American Steel and Wire who had done some preliminary engineering work at Mt. Hood. He thought chairlifts were but mere toys. Hansen, with an eye to business, insisted that while this chairlift wasn't much, it would bring in much needed business. Riblet and Hansen had a disagreement over the chairlift but in the end Riblet gave Hansen a free hand. It was also a watershed moment as Riblet withdrew from the daily operations and Hansen, effectively, ran the business. It also foretold of the future shift in the core activity in Riblet Tramway Company's business.

Hansen had numerous obstacles to overcome to bring the Mt. Hood tram to completion. Firstly, in 1937 the Riblet Tramway Company was bankrupt and deep in debt. Byron Riblet was adrift at the helm, still reeling from the shock of failure of his business, due to events beyond his control. The collapse of world metal prices ended speculation in mining properties. With no speculation, there was no investment or buying of machinery. Effectively, the Riblet Tramway Company was out of business. It had no contracts for several years. Hansen had to retain draughtsmen, draw up proposals, and submit a bid.

On top of this, he had to strain his overstretched finances by raising a performance bond on the project. With little cash, and business partners withdrawing, the situation was precarious. Fortunately, the money was raised and thus Riblet Tramway Company submitted a tender. Riblet was awarded the contract to build the Miracle Mile single chair lift. It was the first chairlift in the US with steel towers, and it was the first "publicly owned and built chairlift" in the US.[13] It also "took two years to build the 4953 foot lift that had a vertical rise of 996.4 feet."[14] Interestingly, it was the longest in North America when finished in November of 1939. Norway's Prince Olav and Princess Martha popularized the lift by dedicating it.

There were many problems with the lift. These included derailment of the cable, jamming of chairs against towers, and severe rime icing difficulties. Its popularity was also a problem. As a single lift, its capacity of people was slow, and resulted in long lines of people waiting to get up the lift. Prices for tickets were 35 cents for a single, 50 cents for a round trip, one dollar for three rides and two dollars for all day.[15] Power was provided by a First World War US Navy diesel which generated electricity for the motors on the tram.

The resort was closed in 1942 due to the new war. It reopened after the war and was very popular. Skiing had a resurgence after the war. People of Scandinavian and Nordic ancestry knew the wonders of winter sports. There were also recent US Veterans, who had served in the US Army Tenth Mountain Division, and who also had discovered the joys of skiing. As a result many new resorts were built. Thus the postwar era saw the advent of the modern ski industry. The US had 55 chairlifts in business by 1955.

Peculiarly, against this trend, Timberline Lodge and lifts were derelict and bankrupt by 1954. Some wanted to burn the lodge and remove it. Fortunately, the US Forest Service issued a contract to a young Richard Kohnstamm and he set about re-invigorating the Lodge, lifts (they were several rope tows too) and ski hill. Kohnstamm succeeded in turning around the resort and more lifts were built. It must be said that he had the inside edge—his European cousin, E. Constam had patented the J-bar lift in Switzerland in 1934 and later the T-bar.[16] By today's standards, the ski industry was quaint, small and undemanding. In the intervening 40 years, lifts became bigger, runs became wider and steeper, equipment became better and people became better skiers. As a result of this growth, Riblet's Miracle Mile chairlift was dismantled in 1962 and replaced with a double chairlift. In time, the crowds became bigger.

Those burgeoning crowds provided the impetus for probably the strangest tramway built it America. While having no connection to Riblet Tramway, the Mt. Hood Ski-Way was a fantastic false start in tram technology and a magnificent failure.

In 1947 the huge crowds on the winding WPA road up to Timberline Lodge was providing a dis-incentive for that resort. Traffic and parking was a nightmare. To move people from the highway to the resort a to and fro tramway was planned. With private money and US Forest Service approval, the Mount Hood Aerial Transportation Company was formed. A tramway would move people from the more accessible land at Government Camp three miles up to the Lodge.

It all seemed so perfect, only a design had to be finalized, and construction could begin. The Roebling Cable Company was interested as was the principle for the local Trailways Bus Company. In the design process, the bus company brass prevailed against the experienced common sense of Roebling. They opted for a suspended bus, whose motor acted as the prime mover, as opposed to a moving cable powered from the terminal station. It was a modification of the locally built Sky-Way log hauling system. This unorthodox approach helped doom the project for the bus had to slow down to clear the support towers, and thus slowed its movement; a fact that reduced the number of passengers hauled per hour.

At the same time, a new road was bulldozed to the Timberline ski area which allowed the traffic to flow more freely. The public realized that they could drive for free and be there faster, rather than park at the highway, and pay money for the privilege of riding a slow moving ski lift. The Ski-Way was a monumental failure and cost its investors dearly. Ignominiously, the lift closed its doors a mere 2 years after opening. A large bull wheel from an outdated lift can be seen a Rhododendron, near the Mt Hood park office.

Riblet, on the other hand, had success with Mt. Hood. That project secured a foothold in the industry. Riblet was soon contracted again by the US Forest Service to build another chairlift at Donner Pass, California. Different again from the Miracle Mile design, the Donner Pass lift had to account for the heavy snows in the Sierra Nevada mountains. To this end the upper and lower terminals were built on gigantic ramps, which would allow for the station structure heights to be adjusted accordingly dependent upon the snow height. (See Appendix V for a listing of modern Riblet Chairlifts.) The Donner Pass chairlift was finished in 1939.

Notes—13

[1] See J. Fahey. "The Brothers Riblet." *Spokane Magazine. November 1980.*

[2] Unpublished biography of Byron Riblet by Carl Hansen.

[3] *Report of the Minister of Mines* as summarized in MINFILE. www.em.gov.bc.ca/cf/minfile for "windpass."

[4] Ibid. p. 2.

[5] Frank Anderson. *The Dewdney Trail.* Frontier Publishing. 1969.

[6] The British retreat from Oregon was not really a grand one. It turns out the major reasons for that territory becoming American include: a snowstorm which delayed David Thompson's arrival at the mouth of Columbia, and allowed the Astorian fur traders to 'claim' it first; the bitter and pointless War of 1812 which made the belligerents more wary of war; the failure of the British to secure Michigan in that war and thus the subsequent establishment of the 49th Parallel, to American benefit, as the demarcation line in Manitoba in the Treaty of 1818, and the lack of political will on the part of the British for unsettled territory far away on the otherside of Cape Horn. The lack of political will was in part due to the cessation of the Napoleonic Conflict in 1815 and its postwar problems of demobilizing fields of soldiers. This period was the time of the Peterloo Riots and the rise of the Luddites. Not only was it a period of high unemployment and high bread prices, it coincided with the "Year of No Summer." In 1815 an Indonesian volcano blew up spewing ash high into the atmosphere. That blanket of dust obscured the sun and hindered the growth of crops. Fewer crops meant higher food prices and social troubles in Britain. Thus, in 1818 a treaty was signed allowing for joint American-British settlment of Oregon.

Joint ownership of Oregon worked well in the first third of the nineteenth century as the European, Native, fur trading and Hawaiian settler populations all worked together to make a community. Unfortunately, at the time in the Eastern US, the population exploded during the era of Jacksonian Democracy and prosperity. However, the bank failures of 1837 forced many failed US farmers west in the 1830's and 1840's. This was one of the reasons behind the American settlers embarking on the Oregon Trail. In Oregon, HBC Chief Factor John McLoughlin, at the time being humanitarian and welcoming and assisting the starving, trail ravaged US settlers (against the Company's orders,) he would only further lessen the ability of Britain to govern the region. By 1846 with so many US settlers in the territory, Britain had effectively let its greater claim on the territory lapse. Thus, a treaty was signed ceding Oregon to the US, and establishing the 49th parallel as the boundary. It also set a precedent for Canadian sovereignty in the West-that of retreat in the face of dubious and greedy American claims. Hence, after the US secured Oregon and Washington, the San Juan Islands, Point Roberts, the Alaska Panhandle, and a larger Panhandle after the Boundary Commission-a century later we are still seeing infractions into Canadian Territory-be it oil rights in the Beaufort Sea or the US dictating internal, domestic Canadian policy as is the case of softwood lumber. Ibid. p. 30.

[7] Ibid. p. 15.

[8] Ibid. p. 30.

[9] Hedley Brochure. Hedley Heritage Museum Society. Hedley, 2000.

[10] The Writers Project provided money to compile histories of each state. They are the first, accurate, detailed guidebooks of each area. While dated now, they still provide indispensable information to the historian.

[11] Robert Trennert. *Riding the High Wire.* p. 100.

[12] Riblet Tramway was contracted to do work on this mining tramway thus the eventual chairlift had a Spokane connection.

[13] J. Grauer. *Timberline and a Century of Skiing on Mt. Hood.* p. 10.

[14] Ibid.

[15] Ibid.

[16] Constam was an innovator and, it turns out, built a very early ski lift at Davos, Switzerland. He was assisted in his efforts by Bleichert of Leipzig. There were other lifts in the Alps prior to this: to and fro tourist trams and carriage funiculars (there were early gondolas at the Aiguille du Midi, France, and at the Matterhorn and Welterhorn in Switzerland.) Much of this work was performed by the Italian firm of Ceretti and Tanfani. There were other tourist trams at Niagara Falls, and Rio de Janeiro as there are to this day. (See Schneigert. *Aerial Tramways and Funicular Railways*. Pergamon Press. 1966 p. 11.)

14

War and its Aftermath

50. Bishop, California

Pine Creek Mine

In the parched, southeastern valleys of California were many mines and tramways. There was a famous Leschen tramway at Cerro Gordo and two others in Death Valley. Culminating the local curiosities was the imponderable Southern Pacific narrow gauge line that connected the region with the outside world.

Numerous mineral deposits were, and still are, in the area. One of these deposits was a body of scheelite, or tungsten ore. During the Second World War the Japanese conquest of Asia had cut off the North American supply of wolfram, the tungsten ore from China and Vietnam. Tungsten was considered a strategic mineral as it is used to harden machine tool cutting devices. So valuable was the mineral that the German Reich had long range submarines carry it from Asia to besieged Germany during the conflict. America's late entry to the war gave it time to coordinate and plan for an involved war. Government agencies were set up to manage and coordinate labour, transport, metals, construction, factory production, oil, and mineral production. The war planning board for mining was called the Metals Reserve.

American Steel and Wire tramway division of United States Steel was contracted in 1940 by the United States Vanadium Corporation to erect a tramway to bring the ore down the steep and exposed mountain at Pine Creek.[1] Gordon Bannerman, American Steel and Wire's chief engineer, supervised the construction. When completed at the end of 1941, "the tramway was 11,000 feet long [and] cost $425,000."[2] The vertical drop was 2900 feet.

The inch and five eighths track cables were split into three sections over the two mile length. The breaks allowed for angle and tension stations to control the descent of the buckets. One inch carrier cables powered the buckets. All were elevated by fifteen wooden towers of various heights. The system used 26 buckets, each with a capacity of 2500 pounds of ore.

The project was a great success, the tram worked flawlessly, the mine supplied the much needed tungsten during the war and remained in operation until 1970. Afterwards a connecting mine tunnel was driven through the mountain.

Tungstar Mine

On the slopes of Tom Mountain, another high grade scheelite deposit was found. The high altitude—12,000 feet—site staked by a Yugoslav named Bill Wasso and his partner G. Crawford in 1937.[3] They sold out to Hollywood money who proceeded to develop the mine in the forbidding mountains near Bishop. Unsure of the mines possible reserves and not wanting to overinvest, the owners were cautious in their building projects. In a word, the project was underbuilt at first.

At such a remote and inaccessible site, the only practicable means of transport was an aerial tramway. In course, Royal Riblet and his Riblet Airline Aerial Tramway Company was contracted to built such a system. Royal had gone into competition against his brother Byron, and his Riblet Tramway Company, though Royal appeared to be competent with tramways, having supervised their installation for 35 years. Royal Riblet did build an unsuccessful tramway for Algoma Steel in Sault Saint Marie, Ontario.

Two Royal Riblet tramways were built at Tom Mountain. A 3200 foot jiback tramway was built to move the ore from the mine workings to a lower level.[4] Where the ore was then transferred to a larger bicable tramway for its journey to the mill. (The 40-ton mill itself was at 7000 foot elevation.) This tram was 10,500 feet long and descended a vertical distance of 4300 feet over that length.[5] It used one-piece inch and one eight track cable and steel towers. All materials were labouriously packed in on mules in the manner of old.

Some problems appeared with the tramway. The tramway towers were too low and thus the heavy snows lying on the ground hampered the buckets travel. And there were more than one dangerous incidence of the tramway cables breaking under the load.

There were many problems with the mill too, breakdowns, and low output. As a result the owners decided to build a new 60 ton mill. That done the original tramway did not have the capacity to keep the mill fed with ore. Thus in 1942 a new tramway was built. It was 11,000 feet long and used wooden towers. It had a 100 ton per day capacity and the descending buckets generated electric power for the mine. Workers also used the buckets to travel to and from the mine until one fell and plummeted to his death.

51. Mouat, Montana

The Second World War created interesting situations in North American mining. Strategic materials were unavailable from their normal overseas supplies due to the actions of the war and thus minerals needed to be located at home. As discussed, tungsten mines were thereby started in California; additionally, mercury mines opened in BC and a chromium mine was put into action in Montana. The reasons for this were that German submarines made ocean shipping perilous in the middle years of the war, effectively separating North America from African mineral supplies. The Japanese occupation of Vietnam and the Philippines in 1942 removed other ready sources for tungsten and chromium—minerals needed in tool and steelproduction.

It should be remembered that during the war the US passed the Mine to Mill Act, which facilitated easier production for mines, the punching of access roads across Federal Land, and the reworking of old mines and railgrades using trucks for the national emergency. The US Federal Government wanted to spur mineral production to counteract the loss of overseas supplies and thus aided the reworking of old mine camps. This law, together with the scrap steel drives, spearheaded much destruction of the Victorian era mining camps, antique machinery, and thus history. At the time, though, the government officials had other concerns.

Chief concerns among the Washington planners were North American sources of raw materials. Fortunately, "a large chromium deposit was found at Mouat, Montana and the United States Government invested $15 million to mine it."[6] Thus in record time the mountain was stripped, drifts driven, a camp, and large mill installed in a move to start production. Anaconda Mining oversaw the mining process under its Anaconda Defense Chrome Operation group formed for the project. There were two drifts—the Benbow and the Mouat—and together these were providing one thousand tons per day of ore. To keep up this production one thousand miners were on task in shifts, working around the clock. Of course, bunkhouses and support structures were also needed to aid such a workforce.

Riblet supplied two tramways to the Anaconda Defense Chrome Operations at Mouat-they were two and one miles long, respectively, and were installed at the Benbow and Mouat pits. Capacity was high: the equipment was built to feed 50 tons of ore per hour. Everything was working in the mine by the summer of 1942 which provided more than half of the US wartime chrome needs.

Despite this large undertaking, and the hurried start up, the mine only produced for 3 months. The fortunes of war had changed by the end of 1942 when other supplies of chrome could be had in North Africa and Sicily after the successful Allied campaigns in 1942. Chromium ore simply was cheaper to produce overseas, and so the Mouat mines were mothballed by 1943. After the war a big producer of chromite was the Balkan peninsula; which is where the discussion leads to next.

52. Yugoslavia

The twentieth century was a tragic period for Yugoslavia. It endured five terrible wars in that relatively short span of time. These wars include the First and Second Balkan Wars, which were fights over the division of the old Ottoman Empire; the First World War which regionally was fought between Austo-Hungary and Serbia; the Second World War, when Germany invaded the independent Monarchy of Yugoslavia; and the terrible Civil War, a war of ethnic rivalries after the breakup of Communist Yugoslavia. All were individually brutal and horrific, yet as a set, can beviewed as longstanding ethnic hatreds, which each generation wanted to act out. Unfortunately, each new war was fought with more modern and deadlier weapons, which brought fresh miseries on the innocent.

For a paper on Riblet, we must briefly mention the Second World War and its aftermath when no fewer than ten German Army divisions occupied the country and attempted to "pacify" it. Obviously, the Yugoslavs took exception to the brutal invaders and fought back in a vicious guerilla war. The German experience in the Balkans was a fateful one—the country was never truly brought under control, and large numbers of troops had to be garrisoned there. In the process of invading Yugoslavia in the Spring of 1941, the action delayed Hitler's invasion of Russia by two valuable summer months. A fact, which prevented his armies advance in Russia by the onset of winter and renewed Russian defense. It can be argued that the Werhmacht's Yugoslavian excursion, and the Russian invasion, allowed for an Allied victory.

Under the leadership of Tito, the Yugoslav guerrillas fought the Germans to a stalemate. In the process, Yugoslavia was destroyed. In 1944 and 1945 the Soviet Red Army drove through Romania and Yugoslavia and defeated the Nazi Armies. Yugoslavia was then within the Soviet Bloc. Wartime planning agreements between the "Big Three" at Tehran and Yalta suggested that Eastern Europe have free electionsafter the war so that each country could decide its government. The arrival of the Red Army changed that agreement. Stalin had no intention of allowing these countries to decide and he would install puppet governments under the control of Moscow. This scenario was not clear to the Western Allies in 1945 as Berlin was divided into zones and the US Army retreated, by agreement, from Czechoslovakia back west of the Elbe River.

Time and again Stalin broke his agreements on free elections, access to Berlin, and other events as they unfolded in Berlin, Czechoslovakia and Yugoslavia. In short, the Iron Curtain had descended and Eastern Europe would be under Stalin's heel.[7] Yet in the three years after the war, the Western Allies wanted to believe that Eastern Europe would have the elections that were stated in the Yalta Agreement. As a show of force, the US Marines sent an expeditionary force to Trieste in 1947 in the postwar readjustment.[8] At the time Yugoslavia was flexing muscle for it wanted to possess Trieste and was subsequently placed under UN mandate.

A Yugoslavian eight Para stamp with an anonymous wood industry cableway.

Considering the wartime costs borne by Yugoslavia, the Western Allies had a favourable attitude toward that Balkan country. Part of this attitude included aid and military assistance. The US wanted to include the ravaged Eastern Europe in its European Recovery Assistance Program, the famed Marshall Plan. In yet another split from the West, Stalin was suspicious of the motives of that program and forbade the 'Eastern Bloc' nations from receiving assistance.

By this time, Tito had emerged as the strong, Communist leader of Yugoslavia—the result of his wartime guerilla leadership. He successfully manoeuvered to keep his position despite the machinations from Moscow and inside Yugoslavia. By 1948, a real power struggle emerged—whether Tito and Yugoslavia would be a lap dog to Moscow, or would be truly independent but still within the

'Confederation' of communist nations. Stalin had effected control earlier on and collectivized the nation's industries. As a result, the national production dropped almost 14%. There were also excesses committed by the occupying Soviet troops. As a result, Tito and Yugoslavia made a public break from Stalin and Moscow. Tito would remain communist and have ties to Russia, but would not be under complete control as Poland, Hungary, East Germany, Romania and Czechoslovakia were. It was a brave position, considering Stalin's purges and, surprisingly, Tito won.

Tito then retained ties to the West. Financial and military aid of over one billion dollars flowed in. Most of it was military aid in the form of surplus war material. Yet the ravaged Yugoslavia also needed trucks, railway engines and carriages, machine tools, steel, road equipment, medicines, mining and farm equipment. The UN Rehabilitation and Relief Administration forwarded $415 million dollars to Yugoslavia immediately after the war in 1945-46. Much of it was for food, clothes and for the immediate living needs of the people. One third of this figure was for industrial rebuilding: this includes money that was sent to facilitate equipment reconstruction at the Trepca lead mines. Interestingly, a Canadian named Lester Pearson was influential in organizing UNRRA affairs in this time period.

After this brief tryst, the courtship between the capitalist and communist nations became intermittent. As stated, international incidents, territory claims, and the topic of corporate compensation hindered a wholesale romance at first. After success with the UNRRA funds and the split with the Soviets the West reconsidered its position to Yugoslavia by 1949. A US State department telegram fromSecretary of State Dean Acheson in abbreviated form explained:

> **Early Sept 1949 Bank approved credits of $20 mil. Of these, $15 mil were to enable Yugo purchase capital equipment and materials to rehabilitate non-ferrous mines and related industries so as permit country, one of the leading producers of bauxite, mercury, copper, lead, zinc and other non-ferrous metals to increase its exports to the US and other hard currency markets.**[9]

Thus the Export-Import Bank of Washington and the International Bank of Reconstruction and Development (IBRD) forwarded the credits. After the Korean War started in the second half of 1950, US Aid was vastly increased to non-Soviet aligned countries.

As a result, between 1949 and 1955 the US Government gave $574 million in economic aid to Yugoslavia.[10] In this manner,

Yugoslavia rebuilt its economy and it became the only strong, independent European Communist country. Part of this US infrastructure investment included aerial tramways for mining and logging. An April 15, 1950 meeting with the Yugoslavs, Export Import Bank, and the US State Department discussed the Yugoslavian balance of payments, loans and economic development:

> Estimated 1950 Dol [lar] balance payments presented (which includes both raw materials and capital equipment) indicates deficit to be met by US of $30 mil after allowing for fol [lowing] loans; a) $2.7 million IBRD timber equipment, b) two already authorized $20 mil Eximbank loans, and c) $12 mil contemplated IBRD loan. [11]

Yugoslavia again had American bank credits to buy US equipment for forestry and for day to day running of the Yugoslav economy.

The International Bank for Reconstruction and Development of Washington authorized three loans to Yugoslavia—the 1950 loan of $2.7 million for forestry as stated in the State Department communique, the 1952 loan of $28 million (terms 25 years at $4\,^1/_2$ percent), and a 1953, $30 million loan.

The 1952 loan was supplied to provide financing for the basic Yugoslav economy.

> The projects include: 1) extension of the electric power facilities; 2) modernization andexpansion of the coal mines; 3) installation of additional equipment at the Bor copper mine, and the erection of a new zinc electrolysis plant at Sabac; 4) erection of new ceramic plant, opening of a new salt mine and expansion of 3 metal working plants, a ceramic plant, a soda ash plant and eight cement plants; 5) erection of new plants for the manufacture of plywood, pulp and kraft paper; 6) acquisition of farm machinery and of equipment for the fishing industry; and 7) importation of equipment for railways and ports. [12]

The Riblet Tramway Company was contracted to build 12 mine and timber carriers for Yugoslavia. The Yugoslavs, it turns out, experimented and developed their own timber cableways in the early 1950's—the Kostnapfel skyline systems.[13] Really these were logging skyline systems that were previously so common in Western North America; the Alpine Europeans discovered them after the war, and many regional engineering works set about trying to perfect a version. Bleichert, Doppelmayr and Pohlig also constructed logging cableways in the Alps.

Despite Herculean attempts to locate information on the cableways in the Balkans, no details have come to light. Where and when the Riblet cableways were employed in that country is unknown. One tramway was at the lead mines of Trepca. It had a cableway that operated for many years and was only scrapped in 1996.

It is known that Yugoslavia had a large mining industry. Copper, bauxite, tin and lead are mined there. There is also an iron industry and the mountains are famed for their lime. Additionally, it is also understood that Yugoslavia leant heavily on its forest resources for reconstruction. In a word, the forests were overcut for lumber to rebuild houses, railways and factories. Lumber was also exported in bulk as a source of income. Some of these cableways were used in logging. The author did acquire an eight Para Yugoslav stamp with an engraving of a cableway at work in the Yugoslavian "Drvna Industrija"or wood industry

Alas, historic investigations at arms length are fraught with problems. Add to this the fact that different languages are involved and a terrible civil war that has destroyed libraries, factories, lives and government offices. Locating information amid this chaos is problematic-particularly when the citizens want to forget the past and get on with just living for today.

53. Albania

Albania has had a peculiar history. Isolated, yet affected by outsiders, it possesses a close proximity to developed Europe and Greece; though even now is prostrate, poor and undeveloped. Geography has affected Albania as well, cut off from its neighbours by the Adriatic sea and high mountains; all have inhibited wholesale development of industry and transportation there. Albania is indeed the unknown and forgotten country of the Balkans.

These themes echo in its recent history with the various currents and counter-currents in the flow of modern events. Many foreign ideologies have visited horrors upon an undeserving Albania. One outcome, no wonder, in Albania is of misguided faith in nebulous structures that come in the forms of Communism, capitalism and a mythic past and fictional future. They serve as an outlet to life in Albania and the lack of performance in Albania of concrete, human organizations.

For much of the last millennia, Albania was under the brutal yoke of the Ottoman Turks. Under their stifling leadership, the country languished. Life changed little from its rural structure, the years were only punctuated by massacres. Early in the twentieth century, the

Ottoman Empire was the 'sick man of Europe.' In turn, the monarchy of Albania came into being after the convulsions of the Balkan Wars (when the Ottomans retreated from the Balkans). Retreating after the Ottoman Empire's defeat in the First World War, the Turks retracted to Asia Minor proper. In the vacuum an opportunistic monarch duly emerged on the throne of the instant kingdom of the Double Headed Eagle.[14] The new country struggled to modernize and recover ground after years of bad rulers and the gains were slow. Such was the state of affairs prior to the deluge of the Second World War.

For in the Depression, across the Adriatic to the West, was Mussolini's Fascist Italy. Mussolini had designs on the Balkan nation and he wanted to incorporate it into his 'Greater Italy;' together with the colonies of Libya, and Ethiopia. As a result, Mussolini sent his armies into Albania in 1939. Three years later, those armies had stalled in a horrific guerilla war in those tall mountains. In turn, the Nazis sent in their legions. Opposing these successive invaders were bands of Communists lead by Enver Hoxha. With Allied support, Hoxha fought a successful war—like Tito in nearby Yugoslavia—and Hoxha emerged victorious in 1944.

Most of the country was destroyed in the war. Despite this fact, Hoxha then set about creating his Socialist People's Democratic Republic. In the immediate aftermath of the war, Albania turned to Yugoslavia for help: joint companies, mining, transport, tariff and economic agreements were signed. Yugoslavia supplied food, technical and development assistance in the bargain. In addition, the United Nations Rehabilitation and Relief fund sent $26 million to assist rebuilding the country. Albanian ties to the west were tenuous and full of ill feeling and so Albania did not garner the aid that Yugoslavia did. After the disastrous war, it seemed that events were improving for Albania. Alas, the murky waters of Balkan politics emerged again.[15]
Tito also had designs on Albania: it was to be an area to harvest profits from. At first, though, Yugoslavia had to invest in its neighbour which it did to the yearly amount of half of Albania's GDP for 1947. When the Albanians became concerned about the nature of the economic development and Yugoslavia's programs, Belgrade excoriated Tirane for its 'revisionist' policies. Albania then proceeded to follow a route towards self sufficiency.

It follows that Yugoslavia was only fattening up the calf before the slaughter—Belgrade was developing Albania in order to incorporate it as a province of Yugoslavia. Tito had earlier annexed the Albanian province of Kosovo into Yugoslavia after the war. (Moscow even supported the Yugoslavian expansion into Albania.) Albania's absorption into Tito's Yugoslavia seemed inevitable, despite what the Albanians thought about the situation. Suddenly, however, these schemes were all reversed, when in 1948 Yugoslavia became non grata in the Communist world. Hoxha kept his job and neck but many other pro-Yugoslav Albanian communists were liquidated in the subsequent purges.

Consequently, Hoxha became paranoid of outsiders and other nations. He had good reason to be for there were other invaders about. In 1949, Britain and the US spearheaded secret, covert invasions using Free Albanian forces in an attempt to overturn the Hoxha regime in Albania itself. These operations continued into the early 1950's but were curtailed after minimal success. One reason for these uprising's failures was their betrayal to the Soviets by the English spy Kim Philby who had direct, detailed knowledge and relayed it to Moscow. As a result, the Albanian insurgents were massacred as soon as they landed inside the Balkans. At the time, Britain and America were concerned about Communist Albania and Yugoslavia, who were aiding Greek communists then fighting a vicious civil war in Greece against Allied backed Monarchists.

In Albania, Hoxa afterwards courted other communist nations following the split from Yugoslavia, but these liaisons went only as far as underwriting his regime. First Russia was exploited and, later, the Chinese were seduced to help pay for a Panglossian, pan-socialist realm. Beyond this Albania retreated into an ultra-Stalinist, xenophobic police state for many years—to the lasting detriment of its citizens.

Now to discuss the topic of industry—in terms here of forestry and mining. About 30 percent of Albania is forested, primarily in the northern part, the Maruras Forest District, and in the Albanian Alps. The proportion of forest cover works out to 2.25 million acres. There are many pine, beech and oak forests there. Even fifty years ago the forests were 'rapidly being destroyed.' The forests had additional meaning for the Albanian economy as wood was used as a domestic fuel, in natural state and as charcoal. It seems that, at this time, the forested regions were high in the mountains, as the woods at lower elevations had already been cut. From this it follows that haulage methods were needed to move the mature trees from where they stood at the ridge level down to the road network in the valley. Hence the resulting need for tramways.

As to tramways, Albania purchased six Riblet tramways to develop their timber industry. Obviously, the trams were bought directly or with Yugoslav help about 1946. Afterwards, US communication and trade with Albania dropped to nil due to the State Department fracas over compensation for nationalized industries.

Albania also has extensive mineral resources. First and foremost are the chromium mines around Pagradec, Klos, Letaj, and Kukes. Copper is also found at: Derven, Bukshize, Erzen, Vele, Narel, Kabash, Klos, Dedaj, Firze, Orosh and Querat.[15] During the Italian occupation, in 1941, iron ore was moved on an Italian aerial tramway at Progradec.[16] Later at another operation tramway machinery was brought in from the Soviet Union to move coal. "The Communist Regime has concentrated in exploiting the Memaliaj Basin where new equipment and an aerial ropeway were installed in the 1950's."[17] Thus a great variety of tramways have operated in Albania.

54. Nalaahu, Hawaii

Mid-twentieth century Hawaii was a different place than it is now. The era of bulk tourism was in the future. Hawaii was then agricultural—the largest ranch in the US is in Hawaii—with pineapples and sugar cane fields dominating the farmland. Throughout history the Islands were also a transit point for ships crossing the vast Pacific. In the ebb and flow of empire, particularity since the Second World War, many US service people visited Hawaii and marvelled at its beauty.

On the southeast side of the big island of Hawaii was the Hutchinson Sugar Plantation. The fields were planted up the volcanic slopes while a refinery, railway and wharf were built at Nahaalu in 1890. Life continued around the pattern of the cane season for decades. The cane was planted, grown and cut. Then a train hauled it to the refinery, where the stalks were boiled down into sugar.

Then, in 1946, a tsunami wrecked the coastal regions of the island: including the town of Hilo, Nahaalu, the Hutchinson Sugar factory, and an occupied school.[18] In the process, the industry on the island of Hawaii was wrecked. The town of Nalaahu was abandoned as was the narrow gauge railway. To resume production a 6000 foot aerial tram was built by Riblet to haul sugar for Hutchinson. Alas, other details are few on the Hutchinson operation. Sugar production has since ceased on the Hawaiian Islands due to world overproduction, competition from sugar beets, and a cane blight.

Notes—14

[1] Joseph Kurtak. *Mine in the Sky*. Publication Consultants. Anchorage, Alaska. p. 65.

[2] Ibid. p. 65.

[3] Ibid. p. 155.

[4] Ibid. p. 157.

[5] Ibid. p. 158.

[6] See www.delamare.unr.edu/deebiog.html#anaconda.

[7] Another school of thought suggests it was the West that promulgated the Cold War. Apparently, Berlin was at first solely to be under Russian control under zones of occupation agreement. Also, the re-militarization of Germany and later creation of the Federal Republic of West Germany incited a negative response from the Soviets. They were mistrustful of the West and the creation of a strong, independent and militarized West Germany. This was the stuff of Russian nightmares considering their recent experience in the war. Reparations were also another issue which were not fulfilled by the Western Powers.

[8] There were aviation incidents over Trieste. US flights from occupied Austria to the Adriatic sometimes strayed over Yugoslavian territory. Subsequently, two planes were shot at in separate incidents causing the deaths of 6 US fliers in a C-47 Dakota.

[9] The full text of the telegram is as follows.

The Secretary of State to the Embassy in Yugoslavia

RESTRICTED PRIORITY Washington, March 1, 1950-4pm.

CABLE BRIEF

EXIMBANK INFORMED YUGOSLAV AMB NOON TODAY RE CREDIT. IT WAS AGREED PROVIDE PRESS 4PM TODAY WITH RELEASE WHICH FOLS FOR THURSDAY MORNING PAPERS;

Early Sept 1949 Bank approved credits of $20 mil. Of these, $15 mil were to enable Yugo purchase capital equipment and materials to rehabilitate non-ferrous mines and related industries so as permit country, one of the leading producers of bauxite, mercury, copper, lead, zinc and other non-ferrous metals to increase its exports to the US and other hard currency markets.
Last Dec and JAN Yugo Govt applied for further Assistance from Eximbank. Present Credit of $20 million is result of Bank's consideration and approval of Yugo applications. Purpose of credit is to enable Yugo to purchase large variety Amer goods, including capital equipment, spare parts, machinery and materials needed to maintain present level of Yugo economy.

The credit, which will be available until March 30, 1951, bears interest at $3^1/_2$% per annum and will be amortized in 14 equal semi-annual installments beginning Jan 1. 1954. A relatively extensive period of grace for repayment of principal has been provided to enable Yugo to meet heavy payments falling due in next 3 years.
-Acheson (Dean Acheson, Secretary of State)
US Council of Foreign Relations. 1950, Vol IV, p. 1378.

[10] J. Lempke. *Yugoslavia as History*. Cambridge University Press. Cambridge. 1996.

[11] The beginning of the telegram is enlightening as it shows that the US is cognoscent of giving all this money away. The Secretary of State Dean Acheson to the Embassy in Yugoslavia (See Foreign Relations. April 15, 1950. Vol. IV. p. 1398.)

[12] The International Bank of Reconstruction and Development, 1945-53. International Bank of Reconstruction and Development. Johns Hopkins. Baltimore. 1954. p. 165.

[13] G. Giordiano. *Logging Cablewa ys*. UN-FAO. Geneva, 1959. p. 27.

[14] Albania had the wonderfully named King Zog, who ruled in the inter-war years until deposed in the Second World War. He then wandered the unemployed monarch circuit of European Capitals. The reality of his reign was despotic and unremarkable.

[15] Albania had strained relations with the West. Hoxha nationalized its industry with out compensation to the prewar owners which affected relations with the US. While two British Royal Navy destroyers were sunk with a large loss of life by mines of unknown origin near Corfu in 1946. Albania denied involvement in the matter, though the case stayed unresolved in the international law courts for decades.

[16] Stauro Skendi. Ed. *Albania*. Frederick Praeger. New York. 1956. p. 186.

[17] The Italians were pioneers in cableways. They built many early cablecars. During the Great War the firm of Ceretti and Tanfani erected hundreds of military cableways on the Isonzo campaign. After that war, these became surplus and were pensioned off to the loggers who used them with great effect. These became the source of the Valtellina system which distinguished itself in Italy. How much of this experience, carried over with the Italians to Albania is not known. Ibid. p. 182.

[18] It seems that tsunamis are quite common in the Pacific. While the 1946 one was quite devastating for Hawaii, the 1964 Alaskan Quake created a wave that remodelled the waterfront of several Vancouver Island towns. In 1917 the German Searaider *Seeadler* was causing havoc to Allied wartime shipping in both the Atlantic and Pacific. The Royal Navy, Imperial Japanese Navy and New Zealand navy all could not catch her. It was a tsunami in the Society Islands that finally wrecked the ship. (See Lowell Thomas' *Count Luckner, The Sea Devil*. Garden City Publ. New York. 1927.) Of course, the port of Pago Pago-now in US Samoan Islands-has a scenic passenger gondola up a nearby volcano. For details on the 1946 Hawaiian wave see the PBS TV, Nova documentary "Killer Wave."

Natural Resources and New Recreations

55. Hollyburn Ridge, Vancouver, B.C.

Hollyburn Ridge is a prominent feature of Vancouver, Canada. It overlooks the city from the northwest, gazing down on English Bay to the south, and the Lions Peaks to the north. The North Shore mountains have played a role in the development of Vancouver. The mountains provide a source of water, recreation, and timber. They also provide a picturesque backdrop for the city.

As a beacon for stressed cityfolk, the North Shore mountains have long called out to the citizens. The forests and beaches drew the Europeans to the region. Summer cabins stretched on the seashore from Ambleside to Horseshoe Bay, and they were used by generations of holidaymakers. Timber on the hillside also drew loggers: in 1908 the McNair and Fraser Lumber company used a walking dudley or cable engine to move their logs down Hollyburn Ridge. Later, they built a log flume to move cedar shingle blocks. Similarly, hikers, mountaineers, and later skiers ventured to the many North Shore peaks. By 1922, the loggers abandoned their mountain camps and cabins and the holiday-makers appropriated them for recreation use. Two Swedes subsequently introduced skiing to the Rodgers Creek region and a new industry was born.

The early skiers were an energetic lot for to reach the mountain, they would board the West Vancouver ferry at Coal Harbour for the three-mile trip to Ambleside. There they would ride a bus to the end of 25th Street. Once there, they would hike the remainder of the way, in winter, and often in the dark, to the 3900-foot level. Staying overnight at the First Lake Cabin allowed these adventurers to ski the next day. Long wooden skis, leather boots and strap bindings added to the fun of the experience. There were no lifts, and the skiers had to walk up the slopes for an (un)controlled ride down.

Later, in the 1940's, Hyde Coville ran a hack service using war surplus trucks to move skiers up to First Lake.[1] This was the vanguard of skiing at the time. To develop Hollyburn further in the postwar boom, a company was formed, Hollyburn Aerial Trams, which installed a lift from the end of the city streets to the ski area. As a result, the Riblet Tramway Company built the passenger lift in 1952.[2] The single chairlift ran one mile in distance starting at the 2900 foot level. It was elegantly named the "Chairlift to the Stars." Twenty-six steel lattice towers raised the cables over head while a 6-cylinder Chrysler engine powered this people-mover. Apparently, the lift broke down often. It ran sporadically for 10 years until 1962. Later, it was partially disassembled with pieces moved to another ski area in the Rockies.

With a view to the recreation possibilities, the area was turned into a provincial park. Cypress Bowl ski area was later established amid the sadly logged forest. Modern Canadian chairlift builders provided the installations for this mountain, and these lifts are still in operation today to assist skiers and reckless snowboarders up to the panoramic peaks.

In a cruel twist of fate, the Riblet Company was, in essence, shut out of business in what previously had been its own backyard. The arrival of WAC Bennett and his Social Credit regime in BC after 1952 created a climate that eventually placed trade tariffs on imported machinery; as a result, regional development diktats perversely excluded a company that had done so much to develop BC industry. Consequently, after this date, few Riblet chairlifts were built in BC. The Riblet Company went on to success in markets elsewhere but the Company's cachet (and the community's collective knowledge of Riblet cableways) diminished in the province.

56. Cassiar Asbestos and the Clinton Creek Mine

In the far north of British Columbia, asbestos was discovered at Cassiar, near Dease Lake. The deposit was on an exposed slope, high atop a mountain ridge. The ore was rich in content which allowed an open cast operation to begin; an aerial tram was built by British Ropeways Engineering in 1955 to haul the material. It was a heavy, industrial tram having a four-wheel travel brackets which suspended the multi-ton ore buckets. The tall, slender support towers indicated tramway designs of a different pedigree. A second tram was later built at Cassiar to increase production—that tram was built by the Interstate Company of the US.

Cassiar Asbestos would open another productive asbestos mine in the Yukon—the Clinton Creek operation. Art Anderson, a native trapper, discovered the large deposit on Porcupine Hill in 1957. Just under a decade later, development work progressed with a road, and a bridge over the Forty Mile River.[3] The open pit mine used Northwest and Bucyrus-Eire shovels, D8 Bulldozers, and Haulpak heavy haul trucks to ship the first fibrous ore by 1968. A tramline hauled the ore from the pit on the hill to the processing plant on the flat ground.

This Riblet tramway was one mile long with a 500 foot rise. "The tramline was constructed by Riblet Tramway Company specifically for this unique installation." And the ore was "delivered at the rate of 300 tons per hour with 70 buckets spaced 180 feet apart."[4] Each bucket had a 1.5 ton capacity and used heavy carriers to transfer such a load. The ore quality was not as high as the Cassiar BC mine, but the volume of the ore body was large. The mine ceased operation in 1979.

57. Nepal

By the late 1950's the world was falling asunder to the Cold War between the US and the USSR. The globe was divided up into spheres of influence and countries were courted to join a side. Developing nations realized that they could extort favours from Moscow and Washington, and play both sides off one another as a country collected aid projects. In a continuation of the "Great Game", so famous from Kipling and the Nineteenth Century, Central Asia became a sparring ground for the competing factions. Treaties were signed, governments were bribed and international politics continued as normal.

At this time the Nepalese government officials started to court the Russians and toy with socialistic policies. "As reported the King [of Nepal] admitted having become committed to receive Soviet aid during his trip to Russia where he received the red carpet treatment."[5] Alarmed, the US Government stepped up the plate as suitor to the Royal Nepalese government and proffered its dowry of consumer durables, planes, plant and equipment.

In 1957, the US Government had initiated an aid program to Nepal, and it was desirous that an aerial tram be built to carry goods from northern India over into Nepal. A request for proposals was started and engineers conducted fieldwork in Asia to more properly assess the engineering needed and task required for the job."A considerable amount of the preliminary design work had to be accomplished during 1958 and early 1959 to properly estimate and submit proposals for the lump sum contract portion" of the contract.[6] Once the Riblet Tramway Company had been issued the contract, draughtsmen finalized the tram designs. Care and attention was given to parts so that no piece was too large, as all had to be carried by humans or animals as there were no roads in tbe Himalayan Mountains. Standardized parts were also made to reduce the complexity of assembly for the local untrained workers. Flexibility was also incorporated into the design to allow the site engineer to adjust to field conditions quickly.

Lead times were also a large consideration as the lag between fabrication of parts, and their arrival in Nepal was up to six months. Part of this was due to "the relatively primitive railway system in Nepal and Northern India over which much of the material for the project would have to be delivered."[7] Transhipment was needed as track widths varied by region—India used standard gauge; Northern India, metre track; and Nepal, narrow gauge.[8] Trucks were also used to haul materials in from Calcutta. The behind the scenes actions of the various governments were revealed in a State Department missive:

> **1) We should avoid trying to outbid Soviets or indulge in recriminations against Nepalese. To react sharply either way would be a mistake.**
>
> **2) I propose inform King and GON [Government of Nepal] officials of my regret their failure to keep me advised on negotiations with USSR and especially decision to establish Soviet Embassy in view our previous understanding this subject. In addition, in view size Soviet aid and limited capacity of Nepalese to absorb foreign aid, I feel Nepal will probably not require any increase in US assistance for some time to come...**
>
> **3) For the time being, we should continue present level of aid and projects without major change in emphasis. We should conclude negotiations on aviation project and, once signed, deliver planes immediately. Similarly, hope there will be tangible evidence of implementation Telecommunications and Ropeway projects before monsoon.**[9]

The tramway was not an outright gift it seems.

Work was started in the summer of 1959 with almost a dozen Americans and 2500 Nepalese contracted on for the job. Materials were hauled through the mountains by hand; this was the same method used for preparing foundations, mixing cement and moving cables. Very heavy machine parts like the gear reduction were also moved by gangs of people. While construction preparations were done, the route survey and tower locations were completed. The work was carried on through the monsoon season, and eventually the project was finished in 1961.

The tram itself was 27 miles long and ran from Hitaura to Kathmandu. While the cables would ascend to 9000 feet before they fell away to 4000 feet at the Nepalese capital. There were 280 towers used to support the lengthy spans of inch and one quarter cable track rope. The moving rope was 3/4 inch. Of course, to reduce the deflection and loading over such a distance, the tram was broken up into seven sections. Electric motors activated the cables. Most of the cargo moved was rice and wheat, hauled in to feed the Nepalese. The tram was also designed to allow some return freight down the mountain. Freight was carried on a timber hook-type platform to make a capacity of 25 tons per hour. Shrinkage or theft must have been a problem as the bags of grain were open to all and sundry on the timber hooks.

A road was later built in the mid 1960's from India to Kathmadu and heavy trucks then moved freight, and Indian culture, into the high Himalaya. After that the tramway began a long decline.

> Though the ropeway has such a huge carrying capacity, it has not been fully operated since 1976 due to the lack of proper repair and maintenance. In fiscal 1991/92 it transported 11,502 tons of goods but ferried only 8000 tons in fiscal 1992/93. While it transported 33,560 tons of goods in fiscal 1996/97, it carried 18,152 tons in 1997/98. During that period, the ropeway charged 34.5 Paisa [currency] per kilogram, whereas, the cost was 47 Paisa in other modes of transport.[10]

Obviously, the ropeway was in business until very recently.

There were other benefits to the Kingdom of Nepal other than cheaper freight rates. As the tram was rated at 22 tons per hour capacity, roughly equivalent to the volume of two Indian sub-continent freight trucks, there was a net savings in fuel. The two trucks would have burned about 250 litres of imported diesel, on the other hand, the tramway was moved by electric motors with the current coming from local, Nepalese sources. "In terms of foreign currency, the country could save more than Rs 175,000 required to import oil each day."[11]

Despite these apparent benefits, the lack of good management, technical training, and repairs reduced the capacity of the ropeway to where it is now not operating. The full capacity of the tramway was never used. There were plans in the 1990's to expand the tramway's use to move cement and garbage around Nepal. However, in 2002 the Nepal Transport Company—the operator of the tram—was disbanded. "Whatever benefits the ropeway offered to the country, the infrastructure worth billions of rupees is waiting to be dismantled and sold for scrap value."[12] And so this was another sad end to a Riblet tramway that had so much potential.

Notes—15

[1] After the war, a lift was built by Columbia Engineering on Grouse Mountain in 1948. Some ascribe this as BC's first ski lift, though this title is contested between it and Red Mountain at Rossland. Another early lift was at Norquay, Alberta.

[2] This information was kindly provided by Bob Tapp–of the Hollyburn Historical Association–in a telephone conversation of August 25, 2003.

[3] Fred Stevens, "The Clinton Mine" *Western Miner*. September 1969.

[4] Ibid. p. 2.

[5] Telegram from the US Embassy of India to the US Department of State. April 28, 1959. *Foreign Relations*, 1958-60. Vol. XV. p. 590.

[6] Tony Sowder. "Nepal Aerial Tramway." ASME publications. New York. 1969.

[7] Ibid. p.1.

[8] Keshab Poudel. "Nepal Ropeway-Hanging in History." *Nepalnews*.com. April 11, 2003.

[9] A large mining tram was planned for Afghanistan in the 1970's in the continuance of this Diplomatic Dance with dollars. A coup and later the Soviet invasion of 1979 ended this scheme. The quotation is by Blake of the US State Department. (See *Foreign Relations*. Op. Cit. p. 590.)

[10] See Poudel, Keshab. "Nepal Ropeway. . . ." *Nepalnews*.com. April 11, 2003.

[11] Ibid. p. 1.

[12] Ibid. p. 1.

Conclusion

The seventy five year span of operation for the Riblet Tramway Company—from its inception in 1897 to 1972—was a period which saw many tramway installations and changes: from the company's establishment and Byron Riblet's first general engineering practice, to his first involvement with the early mining tramways, to his improvements and patents, and longer and larger mining tramways. As we have seen, his tramways follow the pioneer era in British Columbia, Alaska and Washington; thence, Byron Riblet prospers with the industrial expansion in the West in the first decades of the twentieth century.

Included in this scheme were his family, the Cities of Nelson and Spokane, and the Guggenheim businesses. We also saw the building of a legacy—the Arbor Crest house and inventions in Royal Riblet's case, and the Riblet Company and these many tramways in Byron Riblet's life. [It also must be said that this set of tramways is probably not complete; there are conceivably other Riblet constructions out there, unknown and waiting to be documented.]

One could possibly say Byron Riblet's life was incomplete too. The Great Depression very nearly finished the Riblet Tramway Company and it also left lasting marks on Byron Riblet—both financial and psychological. That financial cataclysm forced the company to change, bring in new people and migrate from the mining business. Byron Riblet never really recovered from its effects and half-heartedly carried on until his death following surgery at age 87 in 1952.

The Second World War reinvigorated the company with war contracts; the post war boom continued this trend with additional tram orders. Also, the new skiing industry created a seemingly unlimited demand for chairlifts that the company started building. The Riblet Company spearheaded the rise of the recreational industry as it is they who supplied many of the lifts for the regional ski resorts around North America including Squaw Valley, Aspen and Vail. In this way, the private capital and private profit of the mining centres supported technology which was later applied to the public good.

It is ironical that some of these out of the way mining camps, with heavy machinery and heavy snowfall, were changed from industrial centres to ski paradises. Indeed, the towns of Aspen and Alta rejigged some mining tramways into early ski chairlifts. These isolated communities—which often took days for the pioneers to get to—have been transformed into trendy, chic playgrounds. In BC—Rossland,

Nelson and Kimberly have all become ski centres. Even the rugged Lardeau have been transformed into an Adventure-Ski hotspot. A large chalet exists not far from the site of Ten Mile Camp and the Triune Mine.

In a sense, Byron Riblet's extensive experience, first with the early, unreliable models of mining tramway, then his own products provided the basis to building safe, dependable chairlifts. The Riblet Tramway Company later excelled in chairlift construction starting in the fifties and by the sixties it became the largest builder of chairlifts in the world. At the same time it continued to build mining links, but in lesser numbers. Strangely, the Riblet Company's output was far greater in chairlifts than it was in mining tramways—rough figures supplied by Carl Hansen stated that Riblet Company built about one hundred mining tramways as opposed to four hundred ski chairlifts in this seventy-five year period. Thus, by the 1970's the Riblet Tramway Company had its most prosperous years with established ski areas expanding and other new locations starting. That decade brings us to the modern era and the end of this story.

Considering this history, we can see that Byron Riblet's life and products spurred both mining profits and mountain pastimes for many people; it is an amazing record and one that we should celebrate.

Afterword

During the final stages of preparation for this book word reached the author of the closure of the Riblet Tramway Company in Spokane. A late posting to the internet revealed that Ski Area Management Magazine ran a story in April 2003, it said that Doug Sowder, President of the Riblet Tramway Company, had released information that the company would voluntarily withdraw from business on May 2, 2003. The report went on to state that as the company was only selling one or two fixed grip chairlifts a year, it was not enough to keep the company viable. As a result, it would cease operations. Support and spare parts would be provided in the immediate future for past Riblet chairlift customers.

Competition in the marketplace, changes in design and technology, and a move to detachable, high speed chairlifts had completely changed the scope of the skilift market. Business consolidations and European competition meant that the once proud and dominant Riblet Company was no longer "king of the ski hill."It was a sad day indeed and the final end for the Riblet story.

APPENDICES

APPENDIX I

COASTAL BC TRAMWAYS

Quatsino Sound

Prior to the First World War the Yreka Copper Mine operated on Quatsino Sound. Three Scandinavian locals Nordstrum, Berg, and Everson identified the claim in 1898. Subsequently, a wharf, plant and pelton wheel were installed. An aerial tram was also built to service the mine. Royal Riblet paid a visit to it while consulting for Byron Riblet in 1903. Royal Riblet's slim diary entry accurately attests to his being there: "Wednesday, May 6. Arrived Yreka at 7 p.m. Awfully sick around Cape Scott."[1]

The *Report of the Minister of Mines* explained the Quatsino development:

> **Besides the development of the mine property, the Yreka Copper Company has built the finest wharf on the West Coast; the largest and most up to date ore bunkers, capacity 2500 tons, and aerial tramway 3600 feet, on the Riblet system, and two baby trams, 800 and 400 feet respectively, for transferring from the mines to the main bunkers, and a substantial trestle for the running of ore from the bunkers to the ship. Two trial shipments were made of about 100 tons each, the results being most satisfactory.**[2]

The Quatsino mines only operated for one or two years and then faded away despite the heavy investment. Apparently, there was not enough water in the local creeks to activate the pelton wheel and generate air for the rock drill. Considering the rainfall in the area, something was amiss.

Nearby there was an large incline tram at the Jeune Mine at Alice lake. It had rails and a large snubbing brake similar to the installation at Hedley. Fifty years later another incline was built on Northern Vancouver Island to haul copper ore near Port McNeill. It was installed by Robert McLellan and it copied the technology used at Hedley's Nickel Plate Mine. Of course, copper mining returned to the region with the vast open pit Island Copper mine near Port Hardy.

Indian Chief

In 1918 the Indian Chief claim on Sydney Inlet, north of Tofino, Vancouver Island installed a concentrator and tram to process the copper ore. The claim had been discovered some ten years earlier yet had lain fallow, presumably due to the severe isolation of the site. Access at the time was only by steamer from Victoria to Stewardson Inlet north of Meares Island—and that access to Western Vancouver Island was via the seaward side over the wild Pacific Ocean. The minewas located near the top of a bulbous coastal mountain; thus the tram slashed its attempt at passage, two miles through the overwhelming and verdant coastal forest, down to the beach. The carrier's capacity was 25 tons per hour and it was supplied by Riblet. A mill and wharf, and camp were at the beach. The operation ran only a few years and was well overgrown when it was visited by a hydrographic survey in 1932.

Belmont Surf-Inlet Mines

On Princess Royal Island—midway between Kitimat and the Queen Charlote Islands—is the Surf Inlet waterway. Its ingress is on the seaward side of that striking and exposed island. A lonely mine briefly operated there. It was a gold mine and was in operation just after the First World War. To transport the ore down from the mine to the wharf a simple jigback, reversible tram was built by Riblet. It was 2300 feet long using $1\ 3/8$ inch track rope and $1/2$ inch traction rope. Also, it could carry 20 tons per hour. The mine operated until 1926 and produced around 8 million dollars worth of ore.[3]

---

[1] Royal Riblet's 1903 Diary. Royal Riblet Papers. Cheney Cowles Museum. Spokane, WA.

[2] *Report of the BC Minister of Mines*. 1902. p. H234.

[3] Thanks are due to Ron Smith of Delta for his information on this mine. Also, see LeBourdais. Metals and Men. p.81. Tram details came from The *Report of the BC Minister of Mines*. 1920. p. G40.

APPENDIX II

Patents of Byron Riblet

Byron Riblet patented some two-dozen developments spanning three decades starting at the turn of the century. With these patents, he retained a competitive advantage, which allowed his company to prosper. Most of the patents deal with the mechanical actions of aerial tramways, towers, buckets, loaders, wheels and grips. He did patent some non-tramway ideas such as a light for a gun and an airplane launcher, for example, but these were few in number.

The tram patents and US patent numbers are as follows:

Date	Number	Description
1900	660395	tram bucket and grip
1901	665467	sheave
	685346	derrick
	685387	automatic loader for tramway
	685388	automatic dump bucket
1903	742235	discharge terminal
	742236	grip
	744406	clip
1904	761609	bucket latch
	761610	lower terminal
1905	784647	clip
	806571	air cushion for bucket loader
1906	814622	bucket and rope clip
	878157	bucket hoist conveyor
	913564	conveyor hoist (with A. Leschen and Sons)
1908	878157	cablehoist conveyor
1910	953764	grip for tramway
1916	1195732	aerial truck (with Royal Riblet)
1930	1764465	cablegrip
	1774670	landing and launching ways
1931	1799534	trip device
1934	1981196	bucket wheel with bearings.

APPENDIX III

Other Cable Systems

Coastal B.C.

Doratha Morton Mine

In 1899, a gold mine sprang up about one mile south of Fanny Bay. It was on a ridge—at 2400 feet—on the Beaufort Range. These were the peaks of Vancouver Island above Fanny Bay, a tall range that loomed down. The Fairfield Exploration syndicate was the owner of the operation and they installed a Bleichert type tram to haul ore down 8000 feet to a cyanidation plant constructed at tidewater. One flashy mention in the *Report of the Minister of Mines* marks the mine and then it slips into obscurity.

Money Spinner Mine

Far up stunning Harrison Lake, past the Hotsprings and Sasquatch Inn, is the Lillooet River and a few alpine lakes. At the head of the Harrison Lake was the hamlet of Port Douglas, a Gold Rush era staging post that consisted of a few log cabins. Those prospectors carved a path to facilitate travel to the goldfields via the Lilloet River and Lillooet Lake. After the first gold rush abated, other prospectors continued looking for mining wealth. The Lost Mines of Pitt Lake are a testament to this.

Some sixteen miles from Port Douglas—above St. Mary's Well—and at some 4000 feet lay Fire Lake (or sometimes Tipella Lake). Above that lake was the Money Spinner mine. In 1899 an aerial tram and small concentrator were constructed there. A trail led down to the west side of Harrison Lake where a wharf and bunkhouse were built—a hamlet embellished as Tipella City. The wharf was connected by a company steamer that brought supplies in to the mine. As usual, there is a paucity of detail about the operation other than the mine operated for a few years and then disappeared.

Texada Island

Mining played a large part of the economy of Texada Island, that long sliver in the centre of Georgia Strait. There were several iron mines, a copper operation and a limestone mine located there over the years. According to the *Report of the Minister of Mines* in 1902 the Howe Sound Company developed a mine on the northern end of the island. A Vananda mine used a tram to bring the ore to waiting ships. It has not been determined whether it was an aerial or inclined tram at this site.

Nelson Island

The granite quarry at Quarry Bay removed stone for many years. The Old Post Office in Vancouver, UBC and the Parliament Buildings in Victoria contain some of its products. A short incline was built to transport the dressed stone down to the wharf.

Freil Lake

On the north side of Hotham Sound and Earls Cove is Freil Lake. Logging operators in the 1920's built an incline to haul timber down from the top of the ridge on that peninsula.

Eastern Kootenays

Alice Broughton

Two miles north of Creston was the Alice Broughton Mine. First Staked in 1893, it took ten years to industrialize the operation. Exploratory tunnels were driven hundreds of feet into the mountain and machinery was brought in. It was built in 1904. "During the year a concentrator, with a nominal capacity of 50 tons per day, and connected with the mine by means of a Riblet tramway 5400 feet in length."[4]

Most of the output of the mine was in the first five years under the leadership of the Alice Broughton Mining Co. which removed about 7000 tons of lead-zinc ore. The claims and tailings were reworked again in 1925 and 1949.

North Star and Sullivan Mines

Also near Creston was the North Star Mine. It was owned by the MacKenzie and Mann interests—the great Eastern railway financiers. It had a Bleichert designed tram that was built by Trenton Iron

Works; although the North Star mine was soon forgotten in favour of another local mine. Pat Sullivan was a prospector who scouted the Eastern Kootenays looking for prospects, near the North Star, Sullivan located the lead-zinc orebody that took his name. The Sullivan mine became the most successful mine in BC. It produced ore for almost one hundred years, and finally shut down in 2001.[5] Cominco is doing an admirable reclamation job on the region, though no one knows the long term effects of lead mine leachate into the local rivers.

Zeballos Area

Privateer Mine

When gold was discovered on the northwest coast of Vancouver Island in 1935, it set off a classic rush—one of the last in the old style before the Second World War and the new techniques of mining in the post war period. The Privateer Mine in Spud Valley, some six miles from the beach on Zeballos Inlet on Nootka Sound, turned out to be the most productive mine. It was a large underground operation and it was complete with mill and Spud Valley camp. It formed the nucleus from which other mines radiated.

One of these adjacent claims was the Goldfield Mine. It was just one operation in the patchwork of claims that dotted the steep, damp valleys. Interestingly, there is a fleeting mention in Royal Riblet's draughting file about the Goldfield Mine.[6] (It suggests that there was a tram there and that the Riblets did some work at Nootka Sound, though no supporting information has come to light.)

White Star Mine

If one journeys east over the ridge from the Privateer Mine, one entered a new valley. In this valley was the White Star Mine. Indeed, a tunnel was driven through the ridge to the adjacent valley to connect the White Star Mine to the Privateer workings. In this connection, an aerial tram was built to transport White Star ore from its adit across the valley to the portals of the Privateer mine. It was built just prior to the Second World War. A few cables exist amidst the logging debris. No other information on the White Star Mine could be located. Lastly, a welded, eight-foot diameter bullwheel is on display in Zeballos from a homebuilt tram. It serviced the Man o' War mine which was another mine near Spud Valley.

Homestake Mine

Three valleys east of Spud Valley is the Nomash Valley. A logging road takes you up its western flank from where the Nomash River joins the Zeballos River. Just below treeline was a small gold operation called the Homestake Mine. It had a jigback tram and a cyanide plant. It is relatively complete, aside from the understandable decay and encroaching logging— though, it is guarded by cougars.

Zeballos Iron Mine

In 1962, a Japanese company started mining operations to extract iron ore from the picturesque north side of the Zeballos River. As the mine was several thousand feet up a steep slope, a two car inclined funicular was constructed. The cars passed each other half way up. Robert McLellan was the engineer on the project. In turn, the right of way was cleared of timber, grade prepared, rails were laid and a gearbox and windlass installed. It was a very successful unit and the mine operated for almost two decades. One of the incline cars sits outside the Zeballos Museum.

Ptarmigan Mine

South from Zeballos on Vancouver Island, around Great Central Lake and Della Falls, was the Ptarmigan Mine. It had a bright start and an early demise. A tough prospector named Joe Drinkwater located the copper claim on Big Interior Mountain after he travelled in from Bedwell Sound near present day Tofino. Near Della Lake he noticed gold in the mountains. In fact, the early Spanish explorers purportedly found gold, and the legends of vast seams on Vancouver Island arose. Such was the legend that the name Gold River was given, even though the source of the metal was never found.

Meanwhile, Joe Drinkwater continued hand work on his Della gold claim and built pack roads and simple equipment there as well. While the copper claim on Big Interior Mountain was sold to Rudolf Feilding, the 9th Earl of Denbigh. With his financial backing he made plans to develop the mine. Equipment was landed on the beach at Bedwell Sound and teamsters carried it to the up the mountainous, riverine track. A 5000 foot aerial tramway was planned to connect the Bedwell road with the peak.[7] The materials for the tram were ordered, built and delivered. A concentration plant was also in the works. It was hoped to have the mine process a thousand tons of ore per day, concentrate it and ship it to the Tyee smelter at Ladysmith. The machinery was to be installed in the summer of 1914.

Unfortunately, the First World War broke out in August of that year, flows of capital ceased, and investment reconfigured for the war effort. As a result, the mine never reached full production. Immediately after the war the region became Strathcona Provincial Park, thus closing off mining in the southern end of the park.

Miscellaneous Cable Systems

Dog Mountain

The BC Telephone Company constructed an aerial tram at Dog Mountain, near the Cheam Indian Reserve south of Hope starting in 1955. It was built to service the newly installed microwave relay stations that were linking the country. As the topography of that mountain prevented a road, a tram was built. It is a double rope, single car system. It is still in service today using the original cables. Again Robert McLellan was the engineer on the project.

Boston Bar Car Ferry

Boston Bar is a small town in the Fraser Canyon and thus the Valley is bifurcated by that large river. On the western bank runs the Canadian Pacific Railway, a few ranches, and traditional native territory—including the Stein Valley. To access the west bank from the highway, which is on the east bank, a unique car cable ferry was built. A cage large enough to hold one car was suspended on track cables; traction ropes hauled the cage back and forth over the river. The BC Ministry of Highways constructed it around 1939 and it was use for many years until the 1980's. Afterwards a bridge was built to enable loggers to access the Nahatlatch Valley. The carrier was saved and can be seen in that pretty town today.

University of BC

When Premier John Robson passed the University Act in 1890, he embarked on a program of high schools and university education in the Province. Alas, the best laid plans fall to ruin. The Crisis of 1893 severely curtailed provincial finances and the University plans were shelved. Later, the University was endowed with land on the western tip of Point Grey and a Chancellor was hired in 1908. Grandiose plans were drawn up for a sanitized, bastard-Tudor environs. Finally, the [old] library, science and arts buildings were started.

Alas, the Panic of 1913 and the First World War interrupted these plans yet again. For the duration, the students remained in the crowded buildings on the Fairview Slopes of Vancouver. The unfinished concrete building frames on the endowment lands stood as hollow as Premier Robson's University Act. Only the Great March of 1922 by students convinced the government to finish the project at Point Grey. To do so, the stone and mortar for the project would have to be hauled in.

To move the Nelson Island granite facing stone from a barge, up the vertiginous sea slopes, an aerial tram was constructed.[8] Once atop the hill, a narrow gauge train hauled it to the campus. Ironically, despite possessing Chancellors, Presidents, Registrars, schools of engineering, mining and history, many large libraries and a special collections department, the degreed denizens of the University of BC were unable to explain to the author details about its aerial tram. This simple and essential device—which hauled every stone to the ivory tower in which these enlightened people work and look at daily—is considered so mundane that no retained scholar or court chronicler has remarked upon it.

Cable Ferries

BC has and had some interesting cable ferries. At Needles and Robson on the Arrow Lakes there are a few examples. The Needles ferry started in the 1920's, and several incarnations of vessels have visited this site. Each cable ferry has a winding drum in the hull over which a cable is looped. Either ends of the cable are secured against deadmen or anchors on the riverbank. To travel, the vessel uses the cable wind itself across. The cables keep the vessel on station against the action of the current. Power is supplied by a diesel engine on the ship. The technology was simple, effective and often home grown. The winding drums are building elevator braking mechanisms. The other is located north of Needles and situated to assist the logging traffic. Cable ferries are different than the other type, which are reaction ferries.

Reaction Ferries

One simple, elegant technology used for eons was the reaction ferry. They use the power of the current on a river, and "tack" across the river. A cable is needed to hold the ferry on station. Many have been used over the Canadian and US West. Two are in use on the central Fraser River.

Cable Basket Crossings

In remote areas, crossing sizable rivers in flood on foot is impossible. Rather than having to hike miles to a location or log where one can get across the river, cable crossings have been installed. They consist of a stout track rope from which the metal basket is suspended. The traction cables are like a clothesline and the passenger pulls themselves across manually. The clothesline system is needed to obviate the return of the basket to the opposite bank by someone who is standing on the basketless bank. Otherwise the basket would be stuck on the wrong side of the river with no method of returning it. There are numerous examples of cable crossings around the province: Cheakamus River, West Coast Trail, and Little Fort for example. The first was constructed at Soda Creek one hundred years ago. A century on more and more are being installed as they are very useful.

Volks Marine Railway

The Volks Marine Railway was a peculiar sub-sea railway used to haul holiday makers at Brighton, England. Its railway ran in the inter-tidal zone, parallel to the beach, and was elevated on iron legs to separate the passenger compartment from the tides. Electrically powered, it crossed part of the bay during the Edwardian decade. A similar operation using cables was installed at Saint Malo in France at about the same time.

---

[4] *Report of the Minister of Mines.* 1905. p. G145.
[5] Sullivan had the ill luck of being inundated—he was killed in a mine cave-in, in the Coeur d'Alenes. (See LeBourdais. Metals and Men. p. 53.)
[6] See Royal RIblet's Draughting Index. Cheney Cowles Archives. Spokane.
[7] Lindsay Elms. *"Big Interior Mountain—Mountain and the Miners."* www.memmbers.shaw.ca/beyondnootka/articles/big_interior.html
[8] There is only a passing reference to this tram in G.W. Taylor's *Builders of British Columbia.*

APPENDIX IV

Birch Island, B.C.

Mining experienced yet another boom in the 1950's as a result of the Cold War. The West's demand for uranium created an artificial market for that hitherto unused mineral. In the process, prospectors scoured North America and the world looking for U 238 deposits. The Eldorado Mine on Great Bear Lake in the Northwest Territories supplied much of the uranium oxide for the Manhattan Project. Other prospects were developed near Uranium City on Athabaska Lake in Saskatchewan and Elliot Lake in Ontario. South of the border other mines were developed in Wyoming, Colorado and the Dakotas.

Eighty miles north of Kamloops in the middle of the Thompson River sits Birch Island. It was to become the base camp for a sizable mining operation. For up the mountain was the Rexspar Uranium Metals and Mining Company uranium claim. The property was developed by the capable engineer Franc Joubin in 1950.

The claims were at the 4000 foot level, and the mill was to be built on Birch Island due to availability of flat land. An aerial tram was to be built to connect the two points. Riblet submitted a detailed contract proposal. The tram was to be 10,900 feet long and deliver 50 tons per hour. Its vertical fall was 2700 feet, and thus intended to be self propelled.

The mill was to contain a revolutionary new process of heap leaching for refining the uranium using large pressure towers. In the end, the mine and tramway never left the drawing boards possibly due to market conditions. There was an economic recession in 1956. Whatever the reason, the fish in the Thompson River should be grateful, for toxic runoff from the heap leaching process would have severely affected them. The Eldorado Mine and Elliot Lake are two of Canada's most polluted sites at present and for a few thousand years more.

On the following pages is a letter detailing a Proposed Riblet Tramway for a Uranium Mine and was never built.

PRELIMINARY PROPOSAL FOR REXSPAR URANIUM

May 15, 1956

Rexspar Uranium & Metals Mining Co. Ltd.
Birch Island, B.C.
Canada

Attention: Mr. John W. Scott, Manager

Gentlemen:

We submit herewith our preliminary proposal and specifications for all of the materials for a Riblet Aerial Tramway. Cables, machinery and structural steel herein specified are to be used in constructing an aerial tramway about 10,000 feet long, as described below.

The aerial tramway shall be the type known as a Continuous Bicable System in which the buckets, suspended from trucks and running on a stationary track cable, are also positively gripped to a moving, endless flexible rope, whose function is to move the buckets along the line. The tramway will originate at elevation 3830 feet and will proceed to a southwesterly direction for about 2800 feet, at which place an angle station will be installed. The elevation at the angle station will be about 3334 feet.

From the angle station, the aerial tramway will continue in an northerly direction for about 7700 feet to the point indicated by the small map as M-22. All buckets and the angle station shall be designated so the carriers may pass the angle station will connected to the traction rope.

The aerial tramway will operate as follows: Individual tramway buckets will be loaded by means of a manual chute from the loading terminal bin, after which they will be released so that they can move under the force of gravity into the tramway where they will be connected to the moving traction rope. The buckets will advance at a uniform spacing on the loaded side of the tramway through the angle station and thence to the discharge terminal, the buckets will be detached from the traction rope, will run past the bucket trip and will then continue around the end curve and out of the discharge terminal for their return trip to the loading station. Progress of the empty buckets on the return side of the tramway will correspond with that description for the loaded buckets. While the bucket latch is released in the discharge terminal the bucket tub will overturn and spill its load into the discharge terminal bin.

GENERAL DATA

Location of line . . . . Near Birch Island, BC
Slope Length . . . . 10,900 feet
Fall . . . . 2760 feet
Material to be hauled . . . . Uranium ore
Estimated Weight of ore . . . . 100# per cu. ft.
Capacity . . . . 50 tons per hour
Capacity of each carrier . . . . 1200#
Carrier Spacing . . . . 413 feet
Carriers Required 56
Line Speed . . . . 550 feet per minute
Approximate power produced . . . . 112 H.P

BRIEF SPECIFICATIONS OF MATERIAL TO BE SUPPLIED

1. 11,100 ft. of $1\,1/4''$ diameter smooth coil plow steel track cable.

2. 10,100 ft. of $1\,1/8''$ diameter smooth coil plow steel track cable.

3. 22,000 ft. of $1\,1/8''$ 6 x 19 improved plow steel traction rope.

4. 1680 ft. of $5/8''$ 6 x 19 improved plow steel traction rope.

5. 250 ft. of $1\,5/8''$ 6 x 37 plow steel anchorage rope.

6. 250 ft. of $1\,1/4''$ 6 x 37 plow steel anchorage rope.

7. 65 ft. of $7/8''$ diameter, 6 x 19 hoisting rope for controlling the position of the tension carriage.

8. Three (3) track cable couplings, complete with thimbles and wedges for the $1\,1/4''$ cable.

9. Three (3) track cable couplings, complete with thimbles and wedges for the $1\,1/8''$ cable.

10. The necessary wire rope clips for fastening ropes at the tension blocks, track rope tension machinery, and tension carriage.

BRIEF SPECIFICATIONS OF MATERIAL TO BE SUPPLIED

11. Equipment for ten (10) only breakover type structures for supporting the track cables between terminals, consisting of ten (10) pairs of long radius breakover saddles, with an aggregate length of about 240 linear feet in each size.

12. Two (2) only sets of tower machinery for the necessary single bent, A-frame type, towers. These towers will be used where support is necessary in long spans. Machinery for each consists of two (20 rocking saddles with bases and the necessary post bases for connecting from the structural steel to the concrete.

13. One (1) only tension station, where track cable tension can be adjusted, to be located about mid-way between the angle station and the discharge terminal.

14. Three (3) complete sets of track cable anchorage and tensioning equipment.

15. Six (6) complete sets of approach machinery, for terminals and tension station, consisting of rail tongues, rope tongues, approach saddles and deflecting saddles.

16. Fifty-six (56) only complete double truck, double hanger, double grip tramway buckets.

17. Complete telephone system consisting of three (3) instruments, No. 85 HTL galvanized wire for 2-line circuit, the necessary material for wiring in the terminals, the necessary insulators and pins.

18. One hundred fifty-four (154) only traction rope guide sheaves complete with unit mounting brackets. These sheaves shall be used to support the traction rope on breakover sections, towers, tension stations, loading and discharge terminal approaches.

19. Complete machinery required for angle station, consisting of terminal rails, bucket guide rails, traction rope deflecting sheaves for the external curve and similar deflection sheaves for the internal curve.

20. Discharge terminal machinery consisting of terminal rails, side track switch, attacher and detacher, tension carriage with 8 foot diameter idler sheave, and traction rope tension adjusting winch. A single bucket tripping and dumping frame shall also be included with the discharge terminal equipment.

21. All of the necessary operating equipment for the loading terminal, consisting of terminal rails, side track switch, attacher and detacher, air operated loading chute and the traction rope will be separately described below.

BRIEF SPECIFICATIONS OF MATERIAL TO BE SUPPLIED.

22. One (1) complete traction rope drive and control, located at the loading terminal, and consisting of main shaft with the necessary bearings and thrust bearing; combination brake and grip wheel, equipped with seventy two (72) grip wheel grips and two (2) 10" inch wide brake bands; bevel gear and pinion shaft with bearings and the necessary flexible coupling to the electrical equipment; and two (2) sets of brake equipment consisting of levers and connecting cables.

23. One (1) complete electrical drive consisting of 125 H.P. wound rotor motor, combination primary switch, controller and operating duty resistance, thrustor operated brake, overspeed relay, and the necessary push buttons and limit switches.

24. The necessary special equipment consisting of track cable oiler, wrenches for track cable couplings, handle for operating the traction rope winch, etc.

25. The layout of the final profile and also the necessary assembly, erection, foundation plans are required for field construction of the aerial tramway shall be supplied by the Riblet Tramway Company.

ESTIMATED SHIPPING WEIGHTS OF MATERIALS TO BE FURNISHED

Cables and wire rope . . . . 126,000#
Machinery . . . . 141,000#
Structural Steel and Bolts . . . . 205,000#

GENERAL DESCRIPTION

The aerial tramway will be designed for construction from light structural steel sections, which are shop fabricated for easy field erection. The track cables of the line will be supported upon breakover sections. These sections consist of the necessary vertical panels, braced longitudinally for rigidity, which carry large-radius track cable grooves on cross arms at their top. These structures are placed upon permanent concrete footings. At points near the loading terminal and near the discharge terminal, two towers are provided to support the track cables where the load is relatively light.

The loading terminal, angle station and discharge terminal are to be framed in light steel sections, and besides supporting the terminal rails, our framing is designed to support the walls, roof, and floor of the terminals.

Riblet steel construction consists of light pieces which may be readily moved to their final position and assembled by means of bolts and lockwashers. This system has been used by this company for many years with a high degree of success.

The track cables on the loaded side and also on the return side of the tramway are of the smooth coil type construction. The track cable used on the light side is larger than normal to provide better characteristics in the support of the heavy traction rope loads.

Traction rope sheaves on the line structures and the guide sheaves in the terminals, as well as the trolley sheaves, on which the carriers ride, are each equipped with Timken Roller Bearings and to be of the latest proven design of durable steel.

While these specifications may be brief, we offer our best designs and include with this proposal the advantage of our many years of producing aerial tramways for western North America, as well as many foreign sites.

EXCEPTIONS

This specification includes all materials necessary for a complete 10,800 ft. long Aerial Tramway with the exception of the following material and labor which is not supplied by the Riblet Tramway Co.

Foundation material and labor.

Labor of erecting line structures, terminals and angle station.

Terminal Enclosure material and labor of installation.

Lighting systems in terminals.

Materials and labor used in making track rope anchorage.

Freight, handling and insurance costs incurred in shipping from Spokane, Washington to the tramway site.

Canadian Import Duty, Canadian Sales Tax and Provincial Sales Tax, except where payment of such included in our estimates.

Repairs and replacement of materials injured or lost in transit to the site.

SHIPMENT FROM OUR PLANT

Construction plans, clips, anchorage and approach machinery in from 45 to 120 days, as required by the construction crews. Buckets and terminal machinery in 90 days or longer as desired by the purchaser.

PRICE

Two hundred and sixteen thousand dollars ($216,000.00). Machinery packed for shipment f.o.b our plant at Spokane, Washington, steel f.o.b. Seattle, Washington, and cables f.o.b. Vancouver, BC. Plans to be mailed, prepaid by ourselves, to your Birch Island address.

TERMS

The (10%) percent of the contract price to be paid upon the acceptance of this contract. The balance of the payment to be made as shipments progress, up to the full price, less the 10% initial payment. Payments to be made within thirty days after receipt in your office of invoices, accompanied by original bills of lading or steamers receipts. Price quoted is net with no discounts received.

CONTINGENCY

It is understood that the execution of this proposal on our part is contingent upon possible delays due to strikes, fires, accidents or other reasonable interferences beyond our reasonable control. We hereby respectfully submit the above proposal for your acceptance.

(Signed) *Daniel Lyons*.
Vice President
The Riblet Tramway Company.

APPENDIX V

List of Riblet Ski Chairlifts (1953-1969)

DATE	MOUNTAIN	LOCATION	LENGTH (feet)
1953	Stevens Pass	WA	4900'
1954	Mt. Baker	WA	3500'
1955	Mt. Hood (Ski Bowl)	OR	1800'
	Mt. Majestic	UT	3300'
1956	Stevens Pass #2	WA	2200'
	Timberline	OR	3600'
	White Pass	WA	5100'
	Mt. Spokane	WA	4400'
	Aspen #4	CO	3000'
1957	Sun Valley	ID	4500'
	Mammoth #2	CA	2500'
	China Pk	CA	5700'
	Whiteface #1	NY	6400'
	Whiteface #2	NY	4700'
	Squaw Valley	CA	3200'
1958	Aspen #1	CO	2800'
	Aspen #2	CO	8500'
	White Pass #2	WA	5100'
	Lookout Mtn.	MN	1000'
	Pine Mtn.	MI	2200'
	Big Bromley	VT	5900'
1959	Squaw Valley	CA	4400'
	Aspen #2	CO	5300'
	Aspen #6	CO	1500'
	Red Lodge	MT	3900'
	Mammoth	CA	2700'
	Mt. Majestic	UT	2300'
	Broadmoor	CO	2500'
1960	Stevens Pass #3	WA	1000'
	Mt. Mansfield	VT	6600'
1961	June Mtn.	CA	3000'
	Multorpor #1	OR	3300'
	Mt Bachelor #1	OR	4300'
	Dodge Ridge	CA	3200'
	Brundage Mtn.	ID	5700'
	Alpine Meadow	CA	5600'

	Mt. Spokane #2	WA	4600'
	Indianhead	MI	3200'
1962	Clarksville	MO	1500'
	Magic Mile	OR	5500'
	Vail #4	CO	4700'
	Aspen #3	CO	3200'
	Vail #5	CO	5700'
	Crystal Mtn. #1	WA	3700'
	Crystal Mtn. #2	WA	3500'
	Winter Pk	CO	2600'
	Red Lodge #2	MT	3200'
	Ski Acres	WA	3900'
	Alpine Meadows #2	CA	4100'
	Snow Bowl	MT	5500'
	Pine Knob #1	MI	965'
	Pine Knob #2	MI	960'
	Snow Valley	CA	3300'
	Snow Bowl	AZ	6600'
1963	Thredbo #1	Aus	6100'
	Thredbo #2	Aus	5700'
	Wildcat Mtn	NH	4100'
	Buttermilk #2	CO	3700'
	Spruce Pk	VT	3900'
	Aspen #8	CO	5600'
	Crystal Mtn #3	WA	2800'
	Schweitzer	ID	5500'
	Winter Pk #2	CO	5100'
	Winter Pk #3	CO	3000'
	Pilchuk Pk	WA	3700'
	Boyne	MI	2900'
	Boyne #2	MI	2900'
	June Mtn #2	CA	5100'
	Shanty Ck	MI	3600'
	Aspen #4	CO	3000'
1964	Mt Bachelor #2	OR	3900'
	Big Powderhorn	MI	2200'
	Big Powderhorn #2	MI	1700'
	Dodge Ridge #3	CA	2400'
	Dodge Ridge #4	CA	2000'
	Indianhead	MI	3200'
	Stevens Pass #4	WA	2200'
	Snoqualmie #2	WA	2800'

	Little Switz.	WI	750'
	Bridger	MT	5100'
	Mt. Ashland	OR	3000'
	Shanty Ck.	MI	1700'
	White Pass	WA	3000'
	Vail #1	CO	4000'
	Vail #2	CO	6200'
	Vail #3	CO	3200'
	Camelback	PA	2400'
	Mt Baker #2	WA	2300'
	Alpine Meadows #3	CA	3800'
1965	Bogus Basin #2	ID	3100'
	Bogus Basin #3	ID	4400'
	Twin Peaks	CA	3800'
	Crystal	WA	5100'
	Buttermilk	CO	6000'
	Warm Springs #1	ID	3200'
	Warm Springs #2	ID	6100'
	Purgatory	CO	6100'
	Snowmass	CO	1400'
	Boreal Ridge	CA	2700'
1965	Thunder Mtn.	MI	1800'
	Big Bromley #2	VT	1800'
	Big Bromley #3	VT	3500'
	Wenatchee #1	WA	3500'
	Wenatchee #2	WA	6300'
	Snowmass #2	CO	5500'
	Snowmass #3	CO	3800'
	Mt. Parkway	KY	2300'
	Rushmore Pass	ND	1500'
	Snowqualmie #3	WA	2700'
	Gore Mtn.	NY	3700'
	Grand Mesa	CO	6300'
1966	Multopor	OR	3500'
	Timberline #3	OR	5300'
	Hoodoo Bowl #3	OR	3800'
	Winter Park	CO	1800'
	Sunlight	CO	6900'
	Ski Acres #3	WA	3700'
	Aspen #9	CO	1500'
	Boreal Ridge #4	CA	2400'
	Sugar Bowl #4	CA	2100'

1966	Angel Fire #1	NM	1900'
	Angel Fire #2	NM	2200'
	Angel Fire #5	NM	2200'
	Angel Fire #6	NM	3400'
	China Peak	CA	5400'
	Boyne	MI	1500'
	Whiteface #5	NY	2200'
	Whiteface #6	NY	4800'
	Mt. Reba #1	CA	1500'
	Mt. Reba #5	CA	3200'
	Mt. Hood #1	OR	4400'
	Mt. Hood #3	OR	2200'
	Mt. Bachelor #3	OR	1900'
	Snowmass #4	CO	7500'
	Snowmass #5	CO	7900'
	Ski Acres	WA	1500'
	Schuss Mtn.	MI	1300'
	Mt. Reba #8	CA	3000'
	Alpental #1	WA	4200'
	Alpental #2	WA	2900'
	Alpental #3	WA	1700'
1967	Schuss Mtn #3	MI	1200'
	Snoqualmie #4	WA	2600'
	Wenatchee #3	WA	3700'
	Boyne Mtn.	MI	1900'
	Boyne Mtn. #4	MI	2700'
	Jackson Ski	ID	5200'
	Anthony Lks.	OR	3400'
	Alpine Meadows	CA	2800'
	Natural Bridge Pk	KY	2600'
	Bridger Bowl	MT	7300'
	Boreal Ridge	CA	2600'
	Pilchuck Park	WA	2000'
	Gore Mountain	NY	3000'
	Vail #6	CO	6300'
	Squaw Valley	CA	2100'
	Heavenly Valley	CA	5200'
	Dodge Ridge #5	CA	1500'
	Mt. Reba #6	CA	1800'
	Stevens Pass #5	WA	4400'
	Schuss Mtn	MI	1300'
	Crested Butte	CO	6500'

	Sugar Bowl	CO	960'
	Hoodoo	CO	2300'
	Aspen #3	CO	4900'
	Park City #1	UT	4900'
	Park City #2	UT	6600'
	Park City #3	UT	3200'
1968	Timber Lee	MI	1900'
	Hyak	WA	1400'
	Schweitzer #2	ID	2500'
	Schweitzer #3	ID	1300'
	Schweitzer #4	ID	3100'
1969	Arctic Valley	AK	2200'
	Vail #7	CO	4500'
	Yodelin	WA	1500'
	Snowmass	CO	7100'
	Mt. Hood Meadows #4	OR	2300'
	Jiminy Peak	MA	2700'
	Los Alamos	NM	3100'
	Georgian Peaks #3	ON	1100'
	Timber Lee Hills	MI	1400'
	Boyne Highlnds	MI	1600'
	Boyne Mtn.	MI	2100'
	Winter Park	CO	4700'
	Rib Mountain	WI	3300'
	Snoqualmie #5	WA	1000'
	Mt. Majestic	UT	2700'
	Crested Butte #2	CO	3400'
	Purgatory #2	CO	4100'
	Grey Rock #1	ON	1800'
	Georgian Peaks #1	ON	1800'
	Sandia Peak #2	NM	4000'
	Indianhead	MI	3300'
	Big Powderhorn	MI	2400'

(abridged from the Riblet Report, 1970.)

[Note: most of the locations are US States-i.e. MI Michigan, or CA for California. Aus. for Australia and ON for Ontario.]

APPENDIX VI

Aerial Tramways on the Chilkoot Pass

The Klondike Gold stampeders had their 'choice' of routes to the goldfields of Dawson. The two main paths were landing on the beach at Dyea (another beach just a few miles North of Skagway), up the river to Canyon City, the Scales and over the Chilkoot Pass; while the other was landing at Skagway, up Dead Horse Gulch, and over the White Pass. Other overland methods were tried such as travelling via Edmonton, or in via the Stikine or Nass rivers in BC. Both land routes were more difficult and indirect. Sea passage—the long way round—across the Gulf of Alaska, Dutch Harbor and up the Yukon River was possible, though was ice and weather dependent. Most travellers chose either the Chilkoot or White Pass.

After landing on the muddy tideflat at Dyea, the gold seekers hauled themselves and supplies 13 miles up the rocky wagon track to Sheep Camp. There they would camp and prepare for the most gruelling part of the journey: ascending the Scales in winter, the steep, mountainous and snow covered scree slope to the top of the Chilkoot Pass. There the trail became a four-mile incline to the top of the Chilkoot Pass. Incline was the word as the grade started in at 10 degrees, increasing to 20, and finally a gruelling 30 degree slope at the section known as the Scales.

Not only would one have to summit the Pass but do it many times to bring along 2000 pounds of supplies. A measure enforced by the machine-gun equipped North-West Mounted police to ensure that the gold seeker had enough food and equipment to survive in the harsh north. The top of the Pass demarcated the border between Alaska and a sliver of BC and thus was the suzerainty of Canada. After succeeding with the Scales, the punters had to carry their supplies down the hill to Lake Lindeman and the Yukon Territory, on to Bennett Lake where many built scows to navigate down the river to Whitehorse and eventually the goldfields of Dawson.

Travel by the White Pass was not much easier. When the ships docked at the private wharf Skagway, its owner—Captain Moore—charged the stampeders. This was not the only toll in Skagway. One Jefferson Randolf "Soapy" Smith was the resident gangster who perfected many scams to fleece the unwary travellers in his personal town. Once on the trail the packers soon found that it had been misnamed. It was really a muddy track up a ravine, past boulders on the river bed. It also became known as Dead Horse trail as the trail became impassable to beasts and thus they were left to starve and die. Once over the White Pass, the trail was either snowbound or a muddy bog on the descent to Lake Bennett. Then the goldseekers boated to Dawson as well. As the mountain pass was lower than the Chilkoot, it eventually became the route of the White Pass and Yukon Railway.

Until the railway was built any method of making the difficult transport process better was welcome. Winches, toboggans, and native packers were all tried with limited success on the Chilkoot route. But the immediate solution—in the Spring of 1898—was an aerial tramway to the top of the Scales. First a man named Burns ran a sled—hauled by a gas then later steam winch—up the slope.

The Dyea-Klondike Transportation Company suggested an integrated method of travel to conquer the Chilkoot. First a narrow gauge railway to the foot of the pass, then an aerial tram over mountainous section. Thus, in late 1897, a wharf was built and two 5000 foot steel cables hauled in. Steam engines and a horizontal boiler were also brought in to power the aerial tram. The boilers and engines were located at Canyon City to power an electric dynamo, and a seven mile power line built to power the bottom station at the Scales. It began operation in March, 1898. As competition appeared, only the aerial tram was built and the narrow gauge railway remained unfinished.

It was not a grand tram. No intermediate towers were used, and the cables were supported at the top of the pass. Consequently, the buckets passed several hundred feet into the air. It was a jig-back reversible model, using 2 five-hundred pound capacity buckets. Each trip took about fifteen minutes and cost 5 cents a pound.[10] Meanwhile, the Alaska Railroad and Transport Company also built an aerial tram on chilkoot Pass. A gasoline engine powered its wheels and so eliminated need for a boiler. Low slung to the ground, the buckets passed close to the ground over many small towers. Increased competition—in late Spring 1898—forced the trams to merge operations.

The Chilkoot Railway and Transportation Company raised the last and most elaborate tram. It stretched nine miles from Canyon City over the Pass. The Tacoma based company opted for a Trenton Iron Works model standard mining tram. Construction on it started in October 1897 and crews laboured in wintery conditions to finish the task.

The robust tramway had two sections—reducing the size and deflection on the cables. Tension stations took up the load at half way and kept the 5/8 inch track cables taut in their tower saddles. The usual Trenton Iron Works buckets gripped the traction rope with standard Webber levergrips. A steam engine and boiler at the top of the pass moved the traction rope. Fuel for the boilers was brought up by the tram.

Construction of the Trenton tram carried on through a meningitis outbreak in April, 1898. Yet other tragedies visited the crews too. On April 3, a large snow slide cascaded down and killed the gang constructing the tram towers. Gold stampeders were also smothered in their tents. Over fifty people perished.

Despite the deaths, the company persevered and the tram opened in May, 1898. In the Hegg photograph of the Chilkoot pass, the top towers of the three trams can be seen. The Chilkoot Railway and Transportation tram is on the right, the Alaska Railway and Transportation tram in the centre, and the Dyea Klondike Transport on the left. All the trams operated for one year. In May of 1899, the White Pass and Yukon Railway opened and thus superseding the aerial trams. First the trams merged into one company to meet the threat from the railway, then the railway bought out the trams to reduce competition.

One amusing anecdote on the Yukon cableway access involved a lady traveller. Mae Field was a "bride" of the Yukon gold rush. Her husband brought her there for their honeymoon. They did well in the goldfields. Mae made several trips "outside" back to the Lower 48 States. Returning to Dawson from one of these trips—via the Chilkoot—she was laden with suitcases, steamer trunks, parrot cage and parrot. Mae hiked up the Scales–the steep incline over the Pass. By this time the freight aerial trams were in business so she sent her luggage on via a tram. During "Mae's trip, the cable broke when the tram was halfway up the mountain and it crashed down, carrying a pink parrot riding atop the luggage." Mae explained that when the parrot crashed to earth "for once the bird was speechless. . .and after that the bucket landed right beside me, the parrot screamed, 'What the Hell! Cut it out!.'"[11] Her life was as lively as her travels, she later fell out with her husband and worked the dancehalls and cribs selling l'amour in the North.

---

[10] See the National Park Service Website for the History of the Chilkoot and GoldRush.www.nps.gov/klgo/history_tramway-printer.htm

[11] Detail as to which aerial tram, on the Chilkoot, that ruffled the parrots feather's was not elaborated upon; such extraneous extemporization on elevated links were just not added. It would also seem that Polly wants a salted cracker; profanity spouting parrots reoccur in history. King George V had a profane parrot while Gustafe Flaubert briefly had a difficult parrot, too. (For more on Mae Field consult Lael Morgan. *Good Time Girls*. Whitecap, Vancouver. 1998. p. 112.)

APPENDIX VII

Labour Leaders and Mining

William Haywood

One important labour leader was Big Bill Haywood—a charismatic figure who rose to become the leader of the Western Federation of Miners—the dominant union in the West. His life encapsulates the experience and exertions of miners in the West in the quarter centuries before and after 1900. Born in Utah to Mormons, Haywood grew up fatherless, saw his first gunfight at seven, a black lynching at fifteen and the working face of mine at sixteen. After wheeling rock in Nevada, he changed jobs to a boiler fireman in a lead mine in Utah. [Haywood reminisces recall the awful grey pallor of the poisoned lead mine workers.[12]] He then married his teen sweetheart, bought a ranch to give up mining but was dispossessed off the land without compensation by the Government.

Wandering the West to support his family, Haywood returned to the mines at Silver City, Idaho. There his right hand was crushed when it was caught between a descending car and the side of a shaft[13] In the process, Haywood was taken into the union camp. He rose as an organizer first in Silver City then in Kellogg, Idaho. By 1900 he was elected to the board of the union and led the Western Federation of Miners in their fearful strikes in Colorado. This, too, involved gunplay, soldiers, beatings, murders, blacklists and plots against state officials.

Needless to say, the Establishment pined for Haywood's demise. Indicted for the murder of the Idaho governor, Haywood was implicated as, by this time, he headed the WFM. After a sensational trial in Boise, Haywood was acquitted in a suprise turnaround judgment in 1907. He then went on to found the non-craft based Industrial Workers of the World or "Wobblies" which spearheaded some famous strikes in Washington, California, and Massachusetts. Ten years later, in the First World War, Haywood was imprisoned for two years as an agitator. Following his release, he exiled himself to the Soviet Union where he died in 1928 of prison induced tuberculosis.

Joe Hill

Two other labour martyrs are linked to our stories. Joel Hagglund was a consumptive, Swedish immigrant who came to America to escape the overpopulation of Edwardian Sweden. While odd-jobbing around the American West, he was made aware of the structural and class inequalities in that society. As a result, he was driven into the anarcho-syndicalist camp. As part of adjusting to America, he Anglicized his name to Joe Hill. He also found purpose by becoming a travelling troubadour to the working class. He signed on to Haywood's Industrial Workers of the World and took part in some of their initiatives: strikes for loggers, fruit pickers, and railway workers.

In 1913 Joe Hill helped organize a strike near Yale, BC for itinerant, immigrant workers working on the Canadian Northern Railway in the Fraser Canyon. There were issues over the living conditions in camp. There, he wrote the song "Where the Fraser River Flows." But he did not stay around long enough to finish the affair with the railway workers.

The next year found him working on the tram in Park City, Utah. The copper mines of Utah had long been productive. To further that end Leschen was contracted to build a large aerial tram and terminal in Park City. It was finished in 1902. Thus, in 1914, Joe Hill was corralling ore buckets at Park City. However, there appeared to be more in his life. In 1914 there was an attempted robbery at a grocery store in Salt Lake City and in the ensuing mêlée shots were fired by both the assailants and the owner. In the process, the owner was killed, and themasked bandits escaped.

Into this drama, appeared the fact that Joe Hill ended up with shotgun pellets in him the day after the robbery. He was attended to on the quiet by a doctor. Subsequently, the events in the grocery store were discovered. In the process, Joe Hill was arrested and he, puzzlingly, refused to explain his whereabouts for the night of the robbery or how he received the buckshot. With a naive faith, he claimed he was protecting a friend. Thus by the murder trial, he was presenteas the prime suspect and was convicted. And, in such a case, Joe Hill was sentenced to death.

The case became controversial, as the circumstances of the crime are murky, for he was never positively identified as being at the crime scene. Notwithstanding, it appears that he was involved in the crime. Perhaps, he was funding his pro-union activities by moonlight robberies. People begged for clemency for Joe Hill, and he became a cause *celebre* but he was dispatched to the blue collar Big Rock Candy Mountain by a firing squad. The tale of Joe Hill continues to this day, as his memorable last words were "Don't Waste time in mourning, Organize!"

Ginger Goodwin

Less well known, yet a more clearcut incident of class murder, was Ginger Goodwin. He is Canada's Joe Hill. Another tubercular labourer, Ginger Goodwin was an Englishman who worked in mining camps in the Canadian West. He was involved with the Western Federation of Miners and was on the periphery when an ugly strike broke out at the Dunsmuir Collieries on Vancouver Island in 1913. In that case Attorney-General William Bowser called in the army and the strike broken. By 1916, Goodwin was working at the Consolidated Mining and Smelting Company smelter at Trail. Exempt from the army because of medical reasons, Goodwin unwisely continued his organizing activities in Trail. It must be remembered that this was in the depths of a very bloody war, and Britain had just experienced a crippling shell shortage crisis. Those shell casings were made from copper and zinc, the principle output of the Trail smelter.

In the face of this, Goodwin went forth and organized a strike at Trail in 1917. The strike was settled, though he was forced from town. Now, as a marked man in a time of martial law, Goodwin found his military exemption revoked and he was now eligible for war service. The year 1917 saw more bitter fighting on the Western Front at Passchendaele as the Allies sought victory in a martial meatgrinder. To overcome the shortfall of sacrificed manpower, conscription was enacted. Consequently, by 1918, Goodwin was informed that he was obliged to report to the recruiting board.

The summer of 1918 in BC saw the arrival of the Influenza Epidemic and draft dodgers. By this time Goodwin was with his fellow miners at Cumberland, BC; they had little fondness for the army as they remembered the 1913 Strike. To evade the police, Goodwin hid in the mountains above Cumberland. Alas, that year saw redcapped military police and provincial policemen searching for embusqués. Goodwin was tracked down by a Provincial policeman on the trail to his mountain camp. As he tried to run to the forest, the policeman shot him in the back and killed him.

---

[12] Boyer. *Labor's Untold Struggle*. p. 147.
[13] Ibid. p. 149.

APPENDIX VIII

Colombia

A few words will be said about Colombia. While it is an excursion from the dialogue on Riblet, Colombia was the recipient of some fantastic tramway engineering. As coffee growing was the main industry in Colombia at this time, and as coffee grows at altitude, transport can be difficult. In Colombia the coffee region is near Manzinales which had a railway to connect it to the coast. While the coffee plantations were situated at Mariquita, which were on the eastern side of the Cordillera. "They needed to ship coffee from la zona cafeteria to the Magdalena [River] but the roads through the cordillera were impassible for most of the year."[14] To eliminate mule haulage over washed out tracks a cableway was planned in 1912.

The First World War delayed the construction, as the combatants needed all machinery and steel; though after the war, in 1919, construction began on el cable (as it was called locally.) Apparently, some material was shipped during the conflict as a U-boat sank a cargo load of tram supplies, which further delayed tram construction. The English company Ropeways Ltd. had the contract and supplied the monocable installation. It was 45 miles in length and used 400 towers. The monocable machine traversed ten thousand foot mountains as it carriers sped at four miles an hour. Steam engines activated the carriers. A period, local book exclaimed that the tramway was 'a symbol of progress'.

The tramway was progress. Sacked coffee was moved in volume on freight hooks—10 tons per hour—to the railhead. Passengers and inbound freight could also make the journey. The Manzinales line began operation in 1923 and worked until 1973. By which time Colombia was embroiled in a brutal civil war, ensuring that civil engineering and civil society were cast off into a maelstrom of guns, murder, kidnapping and narcotics.

---

[14] Stephen Smith's book is an enlightening travelogue, family tree and tome on terrorism. He journeys through modern Colombia in search of the path of his grandfather. His grandfather was an engineer for the LaDorada railway, which operated both the trains and the tramway. It seems there was love in the time of tramways. (See Stephen Smith. *Cocaine Train*. Little, Brown, London. 1999. p. 41.)

[15] Original in Spanish. Smith's translation as printed in his book. Ibid. p.41.

APPENDIX IX

Asia

A few words will be said about various cable installations in Asia. While the patterns of tramway development mirror the experience of that in North America, there are a few differences. Indeed, indigenous trams date back to the Middle Ages in Japan—these were really suspension bridges. The first industrial tramway was a Hallidie system imported to work the Ashio copper mines of southern Japan in 1890. In 1901 a Bleichert of Leipzig tram was installed in Japan to move ore and wood. It was very successful and worked for a generation. Cheap local labour costs slowed the need for tram installations, though by the colonial period economic emphasis moved to industrial production.

Japan developed its own aerial trams, under the aegis of the Tamamura Company, and they installed them around Japan and the mines of their Korean colony. The other Japanese colony at this time, Formosa, had a large gold mine and tram at Chui Fang.

Down near the equator at the then British colony of Malaya, the tin cartel installed a large tram at the tin pits; the design was of English lineage—the Breco Company. Another large tram operated at a mine in northern Vietnam during the time of the French mismanagement.

Continental China and Pacific Russia would have used tramways in the mid-twentieth century, though no information has come to light other than a fleeting reference in Royal Riblet's draughting index. Interestingly, the steep Chinese city of Chungking has both an incline and cable car which act like city buses connecting the hilltop, riverbanks, and the Yangtze River in that inland city. Hong Kong has a famous funicular lift but these proliferate around Asia from Vladivostok and beyond.

Australia had two Ropeways installations—one at a tin mine in Queensland at Irvinebank. It now is a museum, while another Ropeways cableway worked at a Tasmanian mine. The Riblet Tramway Company later installed ski chairlifts in the Antipodes at Mount Thredbo. New Zealand had a large coal tram at Stockton. While the Pacific island of Nahru had a large Henderson cableway to move the nitrate salts. The Freeport-McMoran mine still uses a large to and fro tram (built by Bechtel with Canadian assistance) high in the mountains of Western New Guinea.

Apparently, the Riblet Company exported a tram system to Korea sometime in the last half of the twentieth century. It is known that the US Army used tramways as a way to supply and victual its ridgetop firebases during the Korean War. There are also several mines in Korea [an iron mine in Northern Korea had a Bleichert dock loading facility]. Where the Riblet tramway was used in Korea is not known. Nonetheless, It can be seen that tramways were largely imported products into Asia with the various firms competing. Other than the Japanese market, this remained the case with mainly European makers supply the machinery for many years of the twentieth century.

The market has changed considerably in the last few decades. China, India and South Korea have emerged as strong industrial suppliers who excel in building transportation machinery and wire rope. And it is the new market, the leisure industry, rather than the narrow field of mine haulage that trams are being built. For in the mountainous countries from Japan to India and Nepal, passenger cable cars proliferate.

APPENDIX X

Cableways

One important variant of aerial tramways is the cableway. Technically different from the mining aerial tram-cableways only consist of two towers and one span usually over a construction site, much of the engineering is similar. Cableways came into their own around the turn of the twentieth century assisting in the building of many docks, canals, quarries, weirs, and dams. Most of these projects were in the Eastern US.

For a cableway, a heavy cable would be elevated over two wooden towers spanning a construction site. A carriage would be hauled back and forth with a winch, and a fall line would allow for the lifting and placement of supplies. Prior to the age of tall, gantry construction cranes, the cableway was the easiest method of slinging supplies in. Other than a winch, towers, and cable not much else was needed. The Lidgerwood Company of New York pioneered and excelled in building cableways. They erected hundreds of cableways in the opening decades of the twentieth century at every major civil engineering project. The winches and wooden lattice towers became familiar sights on large excavations.

Lidgerwood went on to pioneer the cableway. It developed a long catalogue of towers, cables, carriages, and excavation machinery. Seeing a drawback of the stationary tower, Lidgerwood developed large, travelling pyramidal pylons; these were huge triangles built of steel, and were made mobile on railway tracks. The winch was then housed in the base level of the towers above the wheels. In this way, the whole cableway was movable in a lateral direction, giving the operation 360 degree access to the whole construction site. Masonry, iron, and wood could all be slung into position.

Lidgerwood went on from strength to strength. It used the minor infrastructure projects of the early years to develop the system. It would then come into its own on the Panama Canal project. Each aspect of the canal is fascinating: from the early French failure, to the Yellow Fever and malarial control, to the revolt in the then Venezuelan province of Panama. Each step is intriguing with the huge Bucyrus steam shovels in the Cuelubra cut, to building of locks and lock chambers, and its final completion.

Lidgerwood built massive cableways at the locks to assist in their construction. The towers were 85 feet high over which was saddled 2 $1/2"$ locked coil steel cable.[16] They were used to move wet cement into the concrete forms and haul the steel pieces to the lock gates.

Lidgerwood equipment also improved the productivity of the digging gangs at Panama. Lidgerwood devised a "plow" to unload the dirt spoil from the railway flatcars. Previously, hundreds of men would toil with shovels to empty the cars at the tip. Special flat cars were rigged with a short board back on one side of the car, and over the car couplings sat a sheet of steel plate to make the train into one long, articulated carrier. The steam shovels would load the cars up several feet with dirt and the train would take it away from the cutting. Then the large snowplow like device was hauled by a winch along the cars from the rear to the front. In the process, the dirt was edged off the car. The operation only took a few minutes and compared to the hours of before it was a large improvement.

The Lidgerwood Company is also famous for innovation in logging circles. Steam winches were used to haul logs from the forest for years, though, the heavy logs were always hauled over the ground. Lidgerwood, seeing the success with its overhead cableways, perfected the 'Highlead' system of logging. That is the cables are lifted with a tower or tree and the logs are partially lifted off to ground as they are hauled from the forest. This revolutionized logging when it was first used in the cypress forests of Louisiana in 1904. Soon the method was brought to other forest regions where the high lead and skyline overhead cable systems work to this day.

Lidgerwood crowned its innovation with the building of the Hoover Dam on the Colorado river. Again, it played a supporting role to the Six Companies who built the dam as Lidgerwood erected two huge cableways over the sandstone gorge. The multicable, multi-ton monsters were used to spirit bucket after bucket of cement and to move the penstock pieces onsite. One of the cableways has been kept on standby to help with maintenance of that impressive, groundbreaking dam.

After this, cableways returned to their origin—they went back to dambuilding and obscurity. Most major dams in the world were built with cableways. Name a dam and there was a cableway onsite—Grand Coulee, Bonneville, Dnieper, Italy, Yugoslavia, Switzerland and hundreds of minor, regional dams including the Cleveland in Vancouver'sWater Supply. Cableways were also used on bridge projects such as the million dollar Miles Glacier bridge on the railroad to Kennecott, Alaska.

In Edwardian Canada, when the financiers were attempting to build a ill-conceived third Canadian transcontinental railway, hauling materials to the middle of trackless BC proved to be difficult. Supplies were hauled in by scow on the wild river. At times the crews needed to cross the wide rivers, before the sturdy rail bridges were up. G.W. Taylor comments on the Foley, Stewart and Welch contractor crews building the Grand Trunk and Pacific Railway near Quesnel.

> One of the most difficult jobs was the crossing rivers. On larger streams this was accomplished by building of a boat or barge attached to an overhead trolley powered by a donkey engine on the opposite bank. Such installations cost from $15,000 to $20,000.[17]

Cableways were used in ship construction as well. In the First World War German U-boats were sinking the stock of Allied merchantships. By the end of the war, shipping was in a dangerous plight- To counteract the sinkings, crash building programmes were started for merchantmen. Soon Seattle, Portland, Anacortes and Vancouver, BC sprouted extra shipyards. To minimize capital investment and maximize output, cableways were installed to assist in contruction. Wooden poles were planted along the ways, over which cables were slung. The cableways, acted as cranes to lift the materials to the hulls. Vancouver had many impromptu industrial yards around False Creek which built a series of "War Animal" and "War Camp" ships. Most of thehulls consisted of wood to minimize the use of steel, a strategic and valuable material (as the Allied planners would use ferro-cement as a substitute in the Second World War.) Most of the American ships were not finished by the war's end in 1918, and were summarily burnt or broken up unused.

Lidgerwood had competition in the cableway business—in Britain the Henderson Company of Scotland were famous for their towers and cables on dams and minesites. One anecdote indicates how cable systems have national importance. In 1939, immediately after the outbreak of war, the German U-boats wanted to catch the Royal Navy unprepared as it had done in 1914 when it sank a battleship off the naval base at Scapa Flow. In 1939, U-boat 47 crept into the naval base over an unguarded scholl and sank a capital ship. The Admiralty immediately ordered the filling in of the scholls between the lesser islands to prevent re-occurance of the audacious sinking. To do so Henderson built a large cableway to sling rock to the breakwater and prevent submarines from reappearing in Britain's pre-eminent naval base. Henderson went on to build cableways at quarries, dams and other construction projects around the world.

Again, the cableway solved seemingly difficult problems in the pioneer era. Such ingenuity with timber, cable and winch is to be seen less in the modern day as roads, trucks and crawler equipment allow for heavy machinery to be hauled in intact. Crawler cranes, forklifts and gantry systems have largely replaced the cableway. Only on large spans or dams does the cableway retain its advantage.

---

[16] Ira Bennett. The History of the Panama Canal. 1915.
[17] G. W. Taylor. *The Railway Contractors.* Morriss, Victoria. 1987. p. 63.

Glossary

A quick note should be said about tramway terminology as some terms are confusing, used interchangeably or are not used the world over.

Aerial Tramway - is a large engineered overhead tramway used to move ore, timber or people and the lines often stretch for tens of miles. The term is of North American origin and primarily used in this work.

Ropeway - is a generic term of English origin used to denote funicular railways or overhead aerial tramways. It stems from the earliest days of aerial tramway engineering and, in effect, is a holdover.

Hoist - is a haulage device used to vertically lift cages of men and ore in mines. Specifically and carefully engineered, these hoists are key pieces of mining machinery but separate again from the Aerial Tramways. The wheeled headframes of hoist buildings easily denote mining districts and are often used as a symbol of mining.

Incline - is a railway engineered to ascend and descend a steep hill. Steel cables or manila ropes are used to haul the cars up. This is a North American term, Europeans prefer funicular to denote the same installation.

Funicular - this is the same as Incline, only using the Latin term funiculus or cord instead to indicate a railway on a slope.

Rack Railway - an inclined railway using gears to ascend precipitous slopes. Famous installations are located in New England and Switzerland.

Cableway - a confusing term as it invokes images of ropeway, incline, and aerial term all at once. Yes, it has been used to suggest all three ideas. As a result, in this work, cableway is not a preferred term. In strict nomenclature, a cableway is a shortened aerial tramway only using two towers with a carriage between. They are used to build dams and locks. (Again, confusingly, the term cableway is used interchangeably in Europe to denote a ropeway or aerial tramway.)

Monocable - is a type of aerial tram which only uses one loop of cable both to support and move a carrier. Think of the modern chairlift for an easy example of a monocable tramway.

Bicable - is a lengthy aerial tram, which uses two cables: one trackcable to support the load, and another traction cable to move the load. Most tramways in this book are of bicable design.

Jigback - is a small aerial tram, usually of bicable design in which the car reverses travel at the upper and lower station. Modern passenger gondolas are considered jigbacks (or sometimes reversible tramways).

Cable Car - is a vague term than can describe a gondola used for mountain travel, or a street railway in San Francisco.

Skilift - is a generic term that can describe a ropetow, t-bar, chairlift, or gondola.

Telpherage - is an electrically powered cable car, where the carrierdraws current from hanging cables for the car's motors. It was used primarily for moving mail in English Midland railway stations.

Grip - is a key mechanical device which attaches a carrier to a moving cable.

Angle Station - is a midpoint structure to change alignment of a aerial tramway route. Terminals or Terminal Stations—are loading an unloading point of aerial tramways, usually at different elevations.

Hallidie Tramway - is an early monocable mining tramway built in California by A. Hallidie's Wire rope works and spread around the West after 1867. He held early, key patents, such as the grip wheel.

Bleichert Tramway - a mining bicable tramway built by A. Bleichert and his German Company, or under license in North America by Trenton Iron Works after 1890.

Riblet Tramway - it is bicable mining tramway built by Byron Riblet primarily in Western Canada after 1897. Riblet also built smaller jigback tramways. His company later built chairlifts all around North America.

Leschen Tramway - a bicable mining tramway built by A. Leschen& Sons. Riblet and Leschen Companies formed a business alliance for a decade in the early Twentieth Century. Leschen was also a large cable supplier.

Doppelmayr Lift - is the modern industry leader in cable systems. It is an Austrian company which builds chairlifts, gondolas, and material handling systems the world over. Today the company is a conglomeration of the Bleichert, Von Roll, Garaventa and DoppelmayrEuropean tramway companies and the logical outcome of 150 years of Continental tramway production.

Bibliography

Archives

Nelson Museum and Archives, Nelson BC.
Salmo Library and Archives, Salmo BC.
Nakusp Museum and Archives, Nakusp, BC.
Silverton Museum and Archives, Silverton, BC.
Sandon Museum, Sandon, BC.
Nanaimo District Museum. Nanaimo, BC.
BC Archives and Records Service, Victoria, BC.
Cheney Cowles Museum and Archives. Spokane.
Peck Library, Peck. Idaho.
Stewart Museum, Stewart, BC.
Zeballos Museum, Zeballos, BC.
Alberni Valley Museum and Archives, Port Alberni, BC.
Princton Museum and Archives, Princeton, BC.
Hope Museum, Hope, BC.
Kendrick Hardware and Books, Kendrick ID.
Salmon River Visitor Centre and Museum, Challis, ID.
Park City Historical Society, Park City UT.
Metaline Falls Visitor Center. Metaline Falls, WA.
Wardner Museum, Wardner, ID.
Halfway Museum and Archives, Halfway, OR.
Village of Concrete Library. Concrete, WA.
Geological Survey of Canada Library (Vancouver).
Denver Public Library.
Arthur Lakes Library, Colorado School of Mines.
UBC Special Collections.
Vancouver Public Library Special Collections.

Articles

Aerial Ropeway Eng. Society "The Marmapa Ropeway" *International Ropeway Review.* July 1963.

Attwood, George. "Plant for the Handling and Treatment of Ore at the Silver Cup and Nettie L mines, British Columbia" *Transactions of the Institute of Civil Engineers.* London. 1903. Paper 3486.

Barr, William. "Man Against the Corporation." *Pacific Northwesterner*. 1987.

Beard, R. "Patino Mines and Enterprises at Llallagua, Bolivia." *Engineering and Mining World*. October 1930.

Belknap, R. "Wire Roping the German Submarine" *Scientific American*. March 15, 1919.

Bittner, K. "Milestones in Ropeways History." *Internationale Seilbahn Rundschau*. July 1984.

Buvfers, J. "Historical Survey of Mining in the Ketchikan District Prior to 1952" Alaska State Department of Natural Resources. (Mines and Minerals), 1967.

Carstarphen, Fred. "Truck or Cableway-A comparison of Costs" *Engineering and Mining World*. September 1930.

Dwyer, Charles. "Aerial Tramways in the United States" *OITAF-NACS*. 1988.

Fahey, John. "The Brothers Riblet" *Spokane Magazine*. November 1980.

Frenkiel, Zygmunt. "BRECO celebrates Golden Jubilee." *International Ropeway Review*. January 1966.

Graham, W. "Worlds Longest Bicable Ropeway" *Engineering and Mining Journal*. July 1934.

Guggenheim. Harry. "Building Mining Cities in South America" Engineering and Mining Journal. July 31, 1920.

Hedley Heritage Museum Society. Hedley Brochure. Hedley, 2000.

Hirota, Ritaro. "Ore to Ship by Ropeway in Korea." *Engineering and Mining World*. September 1930.

Hordes. St. "Land Use Investigation, Forest Highway 17." EnvironmentalImprovement Division, State of New Mexico.

Johnson, B "Copper Deposits of the Latouche and Knight Island Districts, Prince William Sound." *USGS*. 1916. B662

Kahn, Edgar. "Andrew Smith Hallidie." *California Historical Association*. June 1940.

Loste, Barbara. "Arbor Crest." Published for Arbor Crest Winery. Spokane. 1995.

Metzger, O. "Aerial Tramways in the Metal Mining Industries." *USGS Publication*. 1937.

Mining and Metallurgical Journal. Tram and Cableway Number. November 5, 1908

Riblet, B.C. "Aerial Tramway Construction in the Andes" *Mining and Metallurgy*. July 1937.

Rich, Beverly. *Mayflower Mill: Reclaimation and Reuse*. San Juan Historical Society.

Roe, Harrison. "Monocable Ropeways." *Engineering and Mining World* April 1930.

RuKeyser, W. "Mining Asbestos In USSR." *Engineering and Mining World*. September 1933

Sowder, Tony. "Nepal Aerial Tramway." *ASME Publications*. New York. 1969.

Stevens, F. "The Clinton Mine" *Western Miner*. September 1969.

"Ropeways and Cableways." *The Sphere*. December 5. 1955,

"Noted Timber Tramway goes into Service Again." *Timberman*. June 1935.

Trennert. R. "Gold Ore and Bat Guano." *Mining History Journal*. Denver. 1997.

Ulrich. J. "Story of Spokane's Famed Makers of Chairlifts is Entrepreneurial Tale ofGreat Idea." *Argus*. March 17, 1967.

Wire Rope Industries. "From 1886 to Expo 86." *Wire Rope News and Sling Technology*. February 1987.

Wuschek. M. "Transportation of Limestone by Means of Aerial Ropeway form the Quarry to the Cement Plant in the Republic of Taiwan." *Materials Handling News*. September 1979.

Books

Affleck, E. *Highgrade and Hotsprings—A History of the Ainsworth Camp.* Ainsworth Hotsprings Historical Society. Ainsworth. 2001.

Bailey, Lynn. *Supplying the Mining World: Mining Equipment Manufacturers of San Francisco 1850-1900.* Tuscon. Western Lore Press 1896.

Barlee, N. L. *West Kootenay: Ghost Town Country.* Canada West Publications, 1984.

Basque, Garnet. *West Kootenay: the Pioneer Years.* Sunfire Publications, Langley. 1990.
_____. *Ghost Towns of Vancouver Island.* Sunfire Publications. Langley, 1993.

Bennett, Ira. *History of the Panama Canal.* 1915.

Boyer. P. et al *The Enduring Vision.* Heath. Lexington. 1995. Vol. 2.

Boyer. B. Labor's *Untold Struggle.* Cameron Assoc. New York. 1955.

Blake, Don. *Valley of the Ghosts—The History Along Highway 31A.* Sandhill. Kelowna. 1988.

_____. *Blakeburn—From Dust to Dust.* Wayside Press. Vernon. 1985.

Blewett, S. *A History of the Washington Water and Power Company 1889-1989.* Washington Water and Power. Spokane. 1989.

Byrnes, R. Ed. *Yugoslavia.* Frederick Praeger. New York. 1957.

Carlson, T. ed. *You Are Asked to Witness-the Sto:Lo in Canada's Pacific Coast Hsitory.* Sto:Lo Heritage Trust. Chilliwack. BC. 2001.

Crowley. T. *Beam Engines.* Shire Publishing. Risborough England. 1996.

Davis, A. Arrow Lakes Historical Society. *Faces of the Past.* Nakusp. 1976.

Dean, F. *Famous Cableways of the World.* F. Muller. London. 1958.

Dorr. J. *Cyanidation and Concentration of Gold and Silver Ores.* McGraw Hill. New York. 1936.

Dubofsky, Melvyn. *We Shall be All—A History of the Industrial Workers of the World.* Quadrangle (NY Times Book Co.), New York, 1969.

Florin, Lambert. *Ghost Towns of the West.* Promentory Press. 1970.

Frontier. *The Dewdney Trail—Hope to Rock Creek.* Frontier Guide. Frontier Publishing. Calgary. 1969.

Frontier. *The Incredible Rogers Pass.* Frontier Publishing. Calgary. 1968.

Gies. J. *Bridges and Men.* Doubleday. New York. 1963.

Giordano, G. *Logging Cableways.* UN-FAO. 1959.

Clara Graham and Angus Davis. *Kootenay Yesterdays, Vol.4.* Alexander Nichols Press. Vancouver. 1976.

Grauer, J. *Mount Hood—A Complete History.* Portland. 1975.

Gower. J. *Fifty Years in Death Valley—Memoirs of Borax Man.* Death Valley 49's. Death Valley. 1969.

Hein, T. *Atomic Farmgirl.* Fulcrum Publishing. Golden. 2001.

Hutchings, Ozzie. *Stewart, BC,* Solitaire, Cobble Hill. 1976

International Bank for Reconstruction and Development. *International Bank for Reconstruction and Development.* Johns Hopkins, Baltimore. 1954.

Jameson, Lone. *Copper Spike.* Alaska Northwest Publishing. Anchorage. 1975.

Johnson, S. *Encyclopaedia of Bridges and Tunnels.* Checkmark. NY. 2002.

Krell, A. *The Devil's Rope—A Cultural History of Barbed Wire.* Reaktion Press. London. 2001.

Kola, Paulin. *Myth of Albania.* NYU Press. New York. 2003.

Kurtak, Joseph. *Mine in the Sky—History of California's Pine Creek Tungsten Mine and the People who were a Part of it.* Publication Consultants. Anchorage. 1998.

Lambert, Florin. *Colorado—Utah Ghost Towns.* Superior Publishing. Seattle, 1971.

Lehman, R. *Bolivia and the United States.* University of Georgia Press. Athens. 1999.

Leier, Mark. *Where the Fraser River Flows—The IWW in BC.* New Star Books, Vancouver. 1990.

Lempke, J. *Yugoslavia as History.* Cambridge University Press. Cambridge. 1996.

Lewty, N. *Across the Columbia Plain.* WSU Press. Pullman. 1995.

Lundberg, Murray. *Fractured Veins and Broken Dreams: Montana Mine andthe WindyArm Stampede.* Pathfinder Publications. Whitehorse. 1996.

May, Dave. *Sandon—The Mining Center of the Silvery Slocan.* Privately Published. 1986.

Mayse, Susan. *The Life and Death of Albert Goodwin.* Harbour. Pender Harbour 1990.

McDuff Leon. *Tererro.* Privately Published. 1993.

Metschler, Charles. *Spokane Street Railways—An Illustrated History.* Inland Empire Railway Historical Society. 1990.

Morgan, Murray. *One Man's Gold Rush.* University of Washington Press. Seattle, 1967.

Morgan, Lael. *Good Time Girls.* Whitecap, Vancouver, 1998.

Mouat, Jeremy. *Roaring Days—Rossland's Mines and the History of BC.* UBC University Press. Vancouver. 1995

Moulton, Candy. *The Grand Encampment—Settling the High Country.* High Plains Press. Wyoming. 1997.
Nelson. Ivar and P. Hart. *Mining Town—The Photo Record of T.*

Barnard and Nellie Stockbridge from the Coeur D'Alenes. University of Washington. Seattle 1984.

Nicholson, George. *Vancouver Island's West Coast.* Moriss, Victoria.1963.Norris, John. Old Silverton, (BC). Silverton Historical Society.-Silverton BC. 1985.

Parent, Milton. *Circle of Silver.* Arrow Lakes Historical Society. Nakusp. 2001.

Peele, Robert. *Mining Engineers Handbook.* Wiley and Sons. New York. (1918, 1927 and 1941 Editions.)

Pellowski, Veronika. *Silver Lead and Hell: the Story of Sandon.* Prospectors Pick Publ. 1992.

Potter, J. *The Life and Death of a Japanese General.* Signet. New York. 1962.

Ramsay. B. *Britannia—The Story of a Mine.* Agency Press. Vancouver. 1967.
_____. *Ghost Towns of British Columbia.* Agency Press. Vancouver. 1963.

Robideau, H. *Flapjacks and Photographs—A History of Mattie Guntermann: camp cook and photographer.* Polestar Press. Vancouver. 1995.

Schneigert. Znignieu. *Aerial Tramways and Funicular Railways.* Pergamon Press. London. 1966.

Schwantes, C. *The Pacific Northwest-an Interpretive History.* University of Nebraska. Lincoln. 1996 ed.

Smallwood, C. et al. *The Cable Car Book.* Bonanza Books. New York. 1980.

Smith, Gibbs. *Joe Hill-the Man and the Myth* Peregrine Smith Books, Salt Lake City, 1969.

Smith. Stephen. *Cocaine Train.* Little, Brown and Co. London. 1999.

Sound Heritage #32. *Where the Lardeau River Flows.* BC Archives. Victoria. 1981.
Sound Heritage. *In Western Mountains-Early Mountaineering in BC.* BC

Archives. Victoria. 1984.

Sparling, Wayne. *Southern Idaho Ghost Towns*. Caxton Printers. Caldwell Idaho. 1996.

Taylor. G. *The Railway Contractors*. Moriss. Victoria. 1986.
_____. *Mining-the History of Mining in BC*. Heritage. Surrey. 1978
_____. *The Builders of BC*. Moriss. Victoria. 1982

Trennert. Robert. *Riding the High Wire-Aerial Mine Tramways in the West*. University Press of Colorado. Boulder. 2001.

Thomas, Lowell. *Count Luckner-the Sea Devil*. Garden City: New York. 1927.

Turner, Robert and Dave Wilkie. *The Skyline Limited*. Sono Nis. Victoria. 1994.

Waite. Donald. *The Fraser Canyon Story*. Hancock House. Surrey.

Wallace-Taylor Alexander. *Aerial or Wire Ropeways-their Construction and Management*. Crosby Lockwood and Sons. London. 1911.

Weis, Norman. *Ghost Towns of the Northwest*. Caxton Printers. Caldwell, 1972.

Whittaker, E. Rossland-*Golden City*. *Rossland Miner*. 1949.

Witcover, J. *Sabotage at Black Tom-Imperial Germany's Secret War in America 1914-1917*. Algonquin. Chapel Hill. 1989.

White, E. and Dave Wilkie. *Shays on the Switchbacks-A History of the Narrow Gauge Leonora and Mount Sicker Railway*. BC Railway Historical Society. Victoria. 1968.

Whyte, J. and T. Brockelbank. *Mining Explained*. Northern Miner with the Southern Magazine and Information Group. Don Mills, On. 1996.

Wolle, Muriel Sibell. *The Bonanza Trail*. Swallow Press. Chicago. 1953.

Woodhouse, Phillip. *Monte Cristo*. Mountaineers. Seattle. 1978.
Woodhouse, Phillip et al. *Discovering Washington's Historic Mines Vol.*

1 and 2. Oso Press. Arlington WA. 1997.

Builders Catalogues

Riblet Tramway Company. Shaw and Borden. Spokane. 1930.

Riblet Tramway Company. *The Riblet Report*. 1970.

A. Leschen and Sons Tramway Company. St. Louis 1930.

William Hewitt. "Wire Rope Tramways it Uses and Applications with special Reference to The Bleichert System." New Jersey 1892.

Roebling and Sons. Circular. c. 1955.

United States Steel. *Aerial Tramways and Light Suspension Bridges* c. 1954.

Ceretti and Tanfani. *Aerial Ropeways*. Ceretti and Tanfani. Milan. 1928.

Lidgerwood Manufacturing Company. *The Lidgerwood Cableway*. Lidgerwood Company.New York. 1907.

Ropeways Limited. Ropeways. London. c. 1918.

Poomong Tea Shoot and Ropeway. Craddock Ropes. Wakefield UK, 1920.

Newspapers

Nelson Daily News
Ferguson Eagle
Trout Lake Topic
Zeballos Privateer
Vancouver Sun
Globe and Mail
Spokane Spokesman Review
Quartz Creek Miner

Private Collections

Robert McLellan, Professional Engineer

Jim Ellis, Professional Engineer

John Fahey, Author

Bill Laux, Historian and tramway worker

John Ledo, Stewart Pioneer

Government Documents

US Government

US Census, Osage, Iowa. 1881.

US Patent Office

US Geological Survey

USGS. *Mineral Resources of Alaska*. 1908. USGS. 1908.

USGS. *Mineral Resources of Alaska*, USGS. 1918.

US Department of War. *Aerial Tramway, Light, Prefabricated M-2* Department of War, Washington. November, 1944.

Foreign Relations

    Eastern Europe, 1950

    Asia, 1957

Washington State

State v. *Northwest Magnesite*.

BC Government

Annual Report of the Minister of Mines. Years 1894-1939.

BC Ministry of Municipal Affairs-Safety Branch. Chairlift Inspection Branch.

Montana

Montana Department of Environmental Quality.

Letters

Chuck Dwyer to author. Aug. 7. 2000.

Sheldon Wimpfen to author. Oct 12, 2000.

Bill Laux to author. Oct 2001.

Jim Ellis to author. Jan. 23, 2002.

Don Blake to author.

Unpublished Documents

The Life and Times of Royal Riblet. Cheney Cowles Museum, Spokane.

Royal Riblet Papers, Cheney Cowles Museum, Spokane.

Hansen, Carl. Biography of Byron Riblet.

Map of Ymir District. Salmo Library and Archives.

Riblet Tramway Company. Preliminary Proposal for the Rexspar Uranium Company. 1956.

Websites

www.stewartbc.ca

www.ghosttowns.com

www.em.bc.gov.ca "MINFILE"

www.riblet.com

www.wikipedia.com

www.doppelmayr.com

www.explorenorth.com/library/yafeatures/bl-granduc1.htm

www.tinja.com

www.fls.org.jm/users/worldeng/andes/andes.html

www.bleichert.de/history.htm

www.inventionfactory.com/history/RHAgen/illumstry.html

www.owensvalleyhistory.com

www.jwmilne.freeservers.com

www.klondikegoldcorp.com

www.vechere.com/smelting

www.his.state.mt.us/departments/ed...rriculum%20Guides/edu_buttearticfinn.ht

www.em.gov.bc.ca/mining/geosurv/minfile/mapareas/beaton.htm

www/argentina4x4.com/norte/nf_cablecarril.htm

www.scoutcampltd2.com.cassiar/the_clinton_mine.html

www.delamare.unr.edu./deebiog.html#anaconda

www.vintners.net/wawine/arbor_crest_birdseye.html

www.plgrove.org/historical/claydump.htm

www.rootsweb.com/~idcuster/custerst.html

www.virtual-crowsnest.ca

www.nps.gov/klgo/history_tramway-printer.html

www.macdonaldbooks.com

www.akbroker.com/history/htm

www.mrsc.org/mc/courts/supreme/o57wn2d/wn2do619.htm

www.atplay.com/hilltop

www.haskell.com

www.hometown.aol.com/Gibson0817/jarbridge

INDEX

Ainsworth, BC, 115
Akun, AK, 172
Alamo Mine, 59
Alaska Boundary, 191
Albania, 258;
  -forestry, 260;
  -mining, 260
Albania-Yugoslav split, 259
Alexandra, BC 24
Algoma, On, 252
Alice Brougham Mine, 282
Alta first chairlift, 223
Alta, UT, 223
American Fork, UT, 223
American Girl Mine, 189
Anaconda Co., 253
Anderson, Art, 268
Anyox, BC, 175,194
Apex Mine, 189
Argentine Mine, 230
ASARCO, 182-185
ASARCO Peru, 231
ASARCo, UT, 223
Ashio Mines, 308
Atlantic Telegraph, 17
Atlin Lake, BC, 167

Balfour, BC, 21
Basques, 224
BC Copper Co., 138
BC Molybdenum Co., 129
BC Silver Mine, 192
Beatrice Mine, 112
Beatson mine, AK, 174
Beatty, NV, 218
Bedaux, Charles, 202
Belmont Surf Mine, 279
Benbow Mine, 253
Benguet Mines, 94
Bessemer, Henry, 18
bicable tramway definition, 82
Big Missouri Mine, 192
bimetallism, 44-45
Bingham Canyon,UT, 184

Black Tom Sabotage, 20
Blakeburn BC, 142-7; disaster, 147
Bleichert Co., 89-90; 94
Bluebell Mine, 29
Bolivia, 8; 229
Bonanza Mine, 175
Bornite Mine, 202
Boston Bar, 285
Braden Mine, 184,187
Braden, Willliam, 187
BRECO, 17; 202; 308 in Gabon, 197
Bre-X scandal, 5
Brill car, 35, 46
Britannia Beach,129-33; flood, 131
Broderick and Bascom Co., 91
Brooklyn Bridge, 20
Brunel, I.K., 16
Bryan, William J., 45
bucket, tramway, 82

California Wire Co., 24,31
Callacuyan, Peru, 232
Camas Prairie RR, 212
Camborne, BC, 109-113
Canadian North Eastern Rwy., 191
Caracoles, 229
Carmen Rosa Mine, 230
Carnation Mine, 66
carriers, 86
Cassiar, BC, 268
  -BRECO tramway, 268
  -Interstate tramway, 268
Cassidy, Butch, 160-1
Catavi, 229
Cayuse Wars, 38
Ceretti and Tanfani Co., 247
Cerro Gordo, CA, 92
chairlift, first skichair, 242
Chewelah, WA, 205
Chile, 187
Chilecito, Argentina, 90
Chilkoot Pass, 19
Chimbote, Peru, 231

333

Clinton Ck. Mine, Yukon, 268
Coalmont Collieries Ltd., 142
Cody, 53
Cody Reco Mine, 53
Coeur d'Alene, 22; 34
Colbert, WA, 39
Colorado River, 217
Colville, H., 267
Comstock Mines, 24
Concrete WA, 209
Conrad Consolidated Mines, 167
Conrad town, 168
Conrad Yukon, 167-71
Conrad, John, 167-71
Consolidated Mining and Smelting Co., 70
continuous type tramway, 82
Cornucopia, OR, 93
CPR Spiral tunnel, 18
Credit Mobilier, 19
Creston, BC, 284
Cripple Creek, Co., 19
Crofton Bay, BC, 127
Cutter, Kirtland, 39
Cypress Bowl Pk., 269

Darrington,WA, 209
de Coccola, Fr., 69
Death Valley, 7
Debs, E., 23
Depression, 237-9
detachable grip, 87
Diamond Match Co., 208
Dog Mtn., 285
Dolly Varden Mine, 175
Donner Pass chairlift, 245
Dopplemayr, 12
Doratha Morton Mine, 281
Doukobors, 22
Drier Bay, 174
Dyea, 301
Dyea Klondike Transport Co., 302

Eagle's Nest Estate,153
Eastman, George, 5
El Portal, CA, 218

Electric Point Mine, 206
Encampment, Wyoming, 158-162
Eva Mine, 109,112
Evening Sun Mine, 189

Fairbanks-Morse oil engines, 76
Fanny Bay, BC, 281
Ferguson, D., 100
Ferguson, BC, 97
Ferris Haggarty Co., 158; 161
Finch Mine, 205
Finlayson tramway, 75
Finning Equipment Co., 188
First Thought Mine, 206
Five Mile Smelter, 106
fixed grip, 87
Freiburg tramway, 26
Freil Lk., 284

Gabon, 197
Garaventa, 12
Goddard, R., 185
Goldfields Mine 283
Gompers, S., 23
Goodwin, Ginger, 306
Granby Co., AK, 173
Granby Mining and Smelting Co., 137
Grand Forks, BC, 137
Grand Trunk and Pacfiic Rwy., 191
Granite Poorman Mine, 115
Great Northern RR, 149
Great Western Railway, 16
Greenwood, BC, 138
grips, 86
Grotto, WA, 210
Guevara, Ernesto, 8
Guggenheims, 181-86, 230-1

Hallidie Cable Car, 27;
- tramway, 26-32, 59, 71, 115
Hallidie, Andrew, 24-9
Hansen, Carl, 238
Hartlieb, 15
Haywood, W., 23
heap leaching, 7
Hearst, William R., 40

Hedley tramway, 240
Hedley, BC, 239-41
Heinze, A., 31
Henderson Cableway, 312
Hewitt Mine, 64
Highland Mine, 115
Highland Valley Copper Mine, 196
Hill, J., 23; 305
Hilo, HI, 261
Himalaya, 267
Hitler, Adolf, 237
Hodgson tramway, 17
Hollyburn Ridge, BC, 267
Homestake Mine, 284
Homestead Act, 4
Hoosac Tunnel, 18
Hoover, Herbert, President, 11
horse whims, 4
Howe Sound Co., 129
Hoxa, Enver, 259-60
Hunter V Mine, 71
Hyder, AK, 191

India, 269
Indian Chief Mine, 279
inventions, 157
Ione, 207-8
Irving, J., 177
Ivanhoe Mine, 57

Jane Basin, 129, 131
Jarbridge, NV, 218
jigback tramway, 87-8
Jumbo Mine, 173; 189;
   -tramway, 189

Kaiser, Henry J., 210
Kathmandu, 270
Keeler, CA, 224
Kendrick, ID, 211
Kennecott, AK, 183-4
Ketchikan AK, 173-4
Kettle river, 137
Kettle Valley Railway, 149
Keystone Mine, 205
King Edward Hotel, 191
King Zog, 263

Kitsault river, 176
Knight Island, Ak, 174
Kohnstamm, Richard, 244
Kostnapfel Skyline system, 257

La Dorada, 196
Ladysmith, BC ,127
Lanark Mine, 97
Lardeau, BC, 97-112
Last Chance Mine, 55, 60
Laurier Mine, 201
Leadville, 19
Lehigh Cement Co., 207
Leonora Mine, 125
LeRoi Mine, 32
Leschen Co., 90-2, 149
Leschen Lever Grip, 141
Leschen tramway, 91-92
Lidgerwood Cableways, 310-1
Limestone Junction, 209
Little Fort, BC, 239
Little Joe Mine, 184
Llallagua, 228
Lone Star Mine, 137-8
Long Hike Mine, 219
Lucky Seven, 189

magnetometer, 8
Mamie claim, 173
Mammoth Mine, 68
Mann, Donald investor, 189
Manzinales, Colombia, 307
Maple Bay Mine, 194
Marshall Plan, 255
Mascot Mine, 240
Maxwell, Robert, 92
Mayview, WA, 211
McBride, Sir R., 23
McKenzie and Mann investors,138
MacKenzie-King, W., 23
McKinley, William,President, 45
Menai Bridge, 16
Metaline Falls, WA, 207
Metals Reserve Board, 252
mine strikes, 22
Mine to Mill Act, 253
Mining Act , 5

Miracle Mile, 243
Molly Gibson Mine, 116-17
Money Spinner Mine, 281
monocable tramway definition, 81
Mont Tremblant, PQ, 242
Monte Cristo, WA, 182
Mouat, Montana, 253
Mount Sicker, BC, 125-8
Mountain Hero Mine, 168-70
Moyie, BC 68-9
Mt. Hood, OR, 242-244
Munday, P., 134
Mussolini, Benito, 259

Nalaahu, HI, 261
Nanaimo tramway, 26
Nelson Island, 282
Nelson Machinery Co., 140
Nepal, 269
Nesbit and Wells Co., 212
Nettie L mine, 102, 105, 106
Newcomen, Thomas, 15
Nickel Plate Mine, 240
nitrates, 229
Nitro-glycerine, 18
No. 1. Mine, 115
No. 7 Mine, 137-8
Noble, Alfred, 17
Noble V Mine, 51-54
Nome, AK, 167
Northern Mine Barrage, 21
Northport, WA, 206
North Star Mine, 282
Northumbria, 7
Northwest Magnesite Co., 200

Oatman, AZ, 217
Odin, 71
Olympic Games, 2
One Big Union, 77
Oregon Question, 239; 246
Oruro, 229
Otto tramway, 97, 119
Ottoman Empire, 258
Owens Valley, CA, 224
Oyster-Criterion, 109-10

Pacific Great Eastern Rwy., 191
Palouse, WA, 33
Panhandle Lumber Co., 208
Panic of 1893, 44-45
Park City, UT, 92
Patino, Simon, 229
Payne Mine, 55
Payne Mtn., 49
Peck ID, 212
Pend Oreille river, 207-8
Peru, 233-34
Pettipiece, Parm, 100, 118
Philby, Kim, 260
Phoenix, BC, 137-8
Pine Ck. Mine, 251
Pioche, NV, 92, 220
Pool Ck., 108
Porter-Idaho Mine, 193-5
Porto Rico Mine, 74
Premier Mine, 192-5; tramway, 188-91; camp, 188
Prince of Wales Island, 173
Prince William Sound, AK, 174
Princess Royal Island, 279
Princeton, BC, 142
Prior, E., 23
Privateer Mine, 283
Pt. McNeill, BC, 240
Pullman, WA, 33
Purkinje, Dr., 17

Quatsino Sound, BC, 278
Quebec, 7
Queen Bess Mine, 57
Quiruvilia, Peru, 232

Rambler Mine, 61
Reco Mine, 56
Red Cliff Mine, 189
Red Mtn., 43
Red Rock Mine, 206
Republic, WA, 206
Rexspar Mine, 290-7
Riblet Airline tramway Co., 156
Riblet, Annie Bell, 38
Riblet, Byron C., (1865-1952)
 children, 37; patents, 282;
 degree, 32; education, 32;
 marriage, 37; office, 38;
 death of, 39
Riblet, Christian, 37
Riblet, family-arrival in America, 37
Riblet-geneaology, 37
Riblet, Hallie, 38
Riblet Leschen Contract, 91
Riblet, Royal, (1871-1960)
 155-7; birth, 155; first marriage, 155; house, 156; work with Riblet Tramway Co., 156
Riblet Tramway Co., 38, 92, 114, 174, 185, 187, 193, 196, 199, 207, 210, 219, 222, 225, 227, 239-40, 245, 259, 270, 271, 275-7, 297
Riblet tramway (first), 51
Riblet, Walter, 38
Riblet, William, 37
Rich Hill Mine, 173
Richmond-Eureka Mine, 57
Robertson tramway, 17
Rock Candy mine, 139
Rock drill, 18
Rockefeller, J., 182
Roe tramway, 17
Roebling, J., 20-21
Roosevelt Mine, 189
Roosevelt, Franklin, President, 237
ropetows, 242
Ropeways Company, 17
Ross Park Railway, 35
Rossland, 32
Ruth Mine, 56
Rhyolite, NV, 218

Salmo, BC, 140
San Elena Mine, 230
San Enrique Mine, 230
Sandon, BC, 49-62
Savery, Thomas, 15
Scapa Flow, UK, 312
Sheep Ck Mine, 140-41
Sherman Silver Purchase Act, 44, 49
shot-creting, 8
Silver Cup Chairlift, 104
Silver Cup Mine, 102-104
Silver Dollar Mine, 111
Silver King Mine, 28-29
Silverton, BC, 64-67
skiing history, 43
Ski-Way tram, 244
Smith, W. Soapy, 6
Snake River tramways, 211
Society Girl Mine, 69
Sovereign Mine, 56
Sowder, Tony, 238
Spokane and Palouse RR, 33
Spokane, WA, 35-38; World's Fair, 37
SS Great Eastern, 16
St Gothard Pass, 17
St. Eugene Mine, 69
Stalin, Joseph, 257
Standard Silver Mine, 67
stations, tramway, 82
Stephenson, George, 16
Steunenburg, F., 23
Stewart, BC, 188-199
Stockton NZ, 308
Sto-Lo nation, 25
Sullivan Mine, 282
Sultzer Mine, 173
Sumas, WA, 209
Sun Valley, ID, 242
Sutter's Mill, 4

Tacoma smelter, 168, 173
Tamamura Co., 308
Teddy Glacier Mine, 113
Tekoa, WA, 34
Telford, Thomas, 16
Teniente, Chile, 187
Tererro, NM, 219-20; accident, 222

Texada Island, BC, 209, 281
Thompson, J., 99
Three Forks, 49
timber carriers-Yugoslav, 257
Timberline Lodge, 243
Tipella Lk., 281
Titanic, 5; 185
Tito, Marshall, 255
Toad Mtn. (Hall), 29-31
towers, the, 84
Treadwell Mine, 177
Trenton Iron Works, 89
Trepca Mines, 256
Trevithick, Richard, 16
Trieste, 255
Triune Mine, 107-109
Tungstar Mine, 252
Tyee Mine, 125-7
typhoid, 43
Tyrell, J., 171

Uncio, 229
Union Pacific RR, 6
University of BC, 285
US-Albania loans, 243-4
US-Yugoslav loans, 259

Valparaiso, 201
Vananda Mine, 209
Vault Mine, 165
Viola Macmillan, 76
Virginia City, NV, 218
Volstead Act, 9; 195
von Alvensleben, A., 41

Waiwatai, WA, 211
Wakefield Mine, 64
Wallace, ID, 34
War Eagle Mine, 32
War of the Pacific, 229
Washington Water and
    Power Co., 36
Watt, James, 15
Western Federation of Miners, 101
White Pass RR, 167
White Star Mine, 283
Whitewater, BC, 49

Wiebe, Adam, 15
Wilde, O., 43
Windfall Affair, 76
Windpass Mine, 239
Windsor Hotel, 100
Windsor, Edward, 197
Windy Arm, Yukon, 167-181
Works Progress Administration
    (WPA), 240
WW II in Yugoslavia, 254-255
Wright, Frank Lloyd, 186

Yankee Girl Mine, 73
Ymir Mine, 71
Ymir, BC, 71-73
Yosemite, CA, 218
Yreka Copper Mine, 279
Yugoslavia, 254-7
Yugoslav-Soviet split, 255-6

Zeballos, 283
Zeballos Iron Mine, 284
Zimmermann telegram, 40
Zincton, Lucky Jim, 61